SECOND EDITION

INTRODUCTION TO
STATISTICS

Mehdi Razzaghi • *Erin Militzer*

Bloomsburg University • Ferris State University

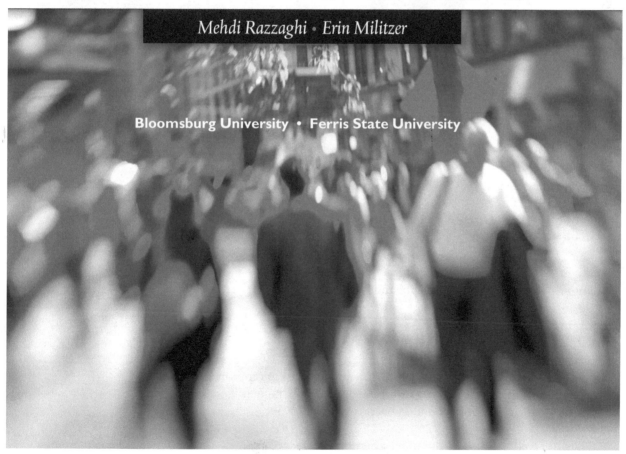

Kendall Hunt
publishing company

Cover image © Shutterstock, Inc.

Kendall Hunt
p u b l i s h i n g   c o m p a n y

www.kendallhunt.com
Send all inquiries to:
4050 Westmark Drive
Dubuque, IA 52004-1840

CONTENTS

Basic Concepts

Section 1: Introduction

When you hear the word 'statistics' you automatically associate it with numbers. You hear about health statistics, education statistics, weather statistics, and so on. You hear about statistics of sports players, government statistics, and many other examples. All these examples point to some numerical information. The fact is that "statistics" is about numbers. We define statistics formally as:

Statistics: Science of assembling, organizing, analyzing, interpreting, and extracting numerical information from data.

Two points are noteworthy about this definition. First that statistics is a 'Science' and it is not a part of any other discipline. Statistics is not a part of mathematics, not a branch of economics and not a division in biology. It is a coherent science for itself. Second that the job of a statistician is not only in one aspect of the process of assembling information, rather the statistician is involved in every step of the process. Today, statisticians are involved in many major research projects and their important role is to ensure that different stages of the project from design and data collection to analysis and interpretation are statistically correct. They work hand in hand with other experts. Statisticians work closely with biologists, physicians, engineers, economists, and with experts from many other disciplines. No quantitative piece of research is considered to be complete without a thorough statistical analysis. The science of statistics has grown tremendously, especially in the last quarter of the century because of its importance and its influence on other disciplines. So, why study statistics? No matter what your major in college and irrespective of what you plan to do in future, you will undoubtedly encounter situations in your life when you have to deal with quantitative analysis. If you read a research paper in your field of study, you will most likely see some statistical terminology. A basic understanding of statistical concepts is considered crucial in critical thinking. It is for this reason that many majors and most departments in universities and colleges require at least one course in statistics as a part of fulfillment of the degree requirements.

Two Branches of Statistics

- **Descriptive Statistics**: Methods to summarize, organize, and hence describe the data using charts, graphs, and numerical measures. In order to present large data sets meaningfully, descriptive methods are applied. Descriptive techniques are powerful and very popular for presentation of statistical summaries.
- **Inferential Statistics**: This is the branch of statistics where we learn how to use the information from the data to make statements about unknown characteristics and unknown features of the entire population. Not only make statements about the population, but also evaluate the reliability of those statements. Then, the question that naturally comes to our mind is how do we measure reliability? Reliability is evaluated using probability.

Here we study both branches of statistics. In fact after introducing the basic definitions in this chapter, in the next chapter we learn methods of descriptive statistics. We then introduce probabilistic concepts and their applications in the next several chapters before we embark on the discussion of statistical inference. Probability is a discipline that has grown significantly in recent years. Clearly, here our goal is not to discuss all aspects of probability theory. Rather, we introduce basic concepts and applications of probability in order to be able to lay a foundation to study statistical inference. We learn concepts of probability, enough to enable us to discuss inferential concepts later in the course. More specifically, we discuss inference in Chapter 2. We then learn about probability and its applications in Chapters 3 to 6 and begin the study of statistical inference in Chapter 7. But first in this chapter, we define basic elements of statistics. In the definition of statistical inference, we described this branch of statistics as being concerned with making statements about unknown characteristics of a <u>population</u>. What do we mean by the word 'population?' What is a statistical population? In the next section we define this and other basic terminologies.

Section 2: Elements of Statistics

Population: A statistical population is defined as the set of all individuals (or items) about whom information is being sought. If we are interested in learning about the opinion of students at a certain college about a certain service such as security, then the population of interest would be the entire set of students in that college. It is clearly very important in any study to have a clear definition of the population of interest. The investigator should have a lucid understanding of who (what) is in the population and who is not. For example, the US government publishes the unemployment figures every month. The population of interest in that case is the entire set of US citizens who are considered to be in the workforce. If a farmer plans to study the yield of a certain crop in the farming lands of the local county, for the farmer, the population under study is the entire farming lands in that county. Similarly, if you plan to learn about the number of hours high school students in your town spend studying outside the school per week, for you the population of interest is the set of all high school students in your town.

Census: The process of collecting information on every individual in the population. In relatively larger populations census is often a very time-consuming process. Census therefore is the process of scrutinizing and collecting information from everyone (every item) in the population. It requires resources and could be very costly. For example, if a manufacturing company wants to collect information on the life of its product, they need to examine every single item produced by the company. In some cases census is even not possible. If we want to gather information on the population of whale in the Pacific Ocean, it would require catching all the whales in that ocean which, of course, cannot be done. For this reason census is not very often practiced and we seldom collect information on the entire population. The US government conducts the national census every 10 years. In practice, we often resort to collecting information on a handful of individuals or items called a *sample*.

Sample: A sample is a subset of the population. Therefore, rather than a census, we collect information on a subset of the population and based on our findings, we try to draw conclusions about the entire population. This is the essence of statistical inference. However, if we want to have reliable outcome and if we want the results of the study to be useful, we must make sure that our sample is a good representative sample. A representative sample would have characteristics that are similar to those of the population. One way to draw a representative sample is using a *random* sample.

Random Samples: A random sample is a sample drawn from a population in such a way that no bias enters the process of sampling, i.e., it is free of any kind of bias. The question is how one can, in practice, draw a random sample. Methods of random sampling are a big topic in statistics and a full discussion of that topic is not within the goals of this course. However, here we briefly mention two methods of random sampling that are used very often for drawing representative samples from populations.

Methods of Random Sampling

1. <u>Simple Random Sample</u>: A simple random sample of size n is a sample drawn in such a way that every group of n individuals would have the same chance of being selected in the sample. If we want to select a sample of four students from a class for a special project and select these four students in such a way that every collection of four students has the same chance of becoming the sample, then we have a simple random sample of size 4. The easiest example of a simple random sample is drawing names out of a hat. Thus if we want to draw a random sample of size 4 from a class, we can write all names on identical pieces of paper and thoroughly mix the papers and select four out a hat. When population is of relatively larger size, we use a computer to draw a simple random sample. Today, many computer programs have a special function to generate random numbers and can be used in simple random sample generation.

2. <u>Stratified Random Sample</u>: Although a simple random sample generally provides a good representative sample from the population and is often used in practice, one drawback is that it may not provide representation from all sectors of the population. For example, if we select a simple random sample of size 50 students from a university, not all ethnic minorities may have representation, or we may not have representation from every year of study. In stratified random sampling, we first divide the population into several subpopulations (strata) according to some criteria. We then select a simple random sample from each stratum (subpopulation). Thus, if we want to sample 50 students and want to make sure that we have representation from every ethnic background, we can stratify by ethnicity, i.e., divide the student population in several groups according to their ethnic background and then select a simple random sample from each ethnic group. Similarly, if we want to make sure that we have representation from every year of study, we can stratify by year of study. As another example, if you want to do a survey of electricity consumption per household in your community, because the amount of electricity consumption varies substantially by the type of residence, e.g., single unit house, condominium, apartment, it would probably be better to stratify by the type of housing by first dividing the population of homes in your community into strata based on the type of housing and then select a simple random sample from each stratum.

Methods of Non-Random Sampling

We ought to mention that *non-random* sampling is not uncommon and is used quite often for collecting information. We mention three methods of non-random sampling as follows:

1. <u>Systematic Sampling</u>—When sampling units are selected on some kind systematic procedure. For example, to select four students from a class for a special task, we can say that we pick every ninth name. This method generally does not provide a good representative sample; however, if we modify it and use what is called '1 in k systematic' sample, then the resulting sample is often reasonably representative. In '1 in k systematic' sampling, we first pick a value of k and from among the first k individuals on the list, we pick one at random. Then, from then on, every kth individual is selected. This method works quite well as long as there are no repeating patterns in the population.

2. <u>Volunteer Sampling</u>—This is when the respondent responds on a completely voluntary basis. Hospitals, different companies, restaurants, and other businesses often use this method to collect information about customer satisfaction and ways they can adjust their policies. The method is used in most surveys and generally suffers from non-response bias. In many cases, in order to encourage better response, some incentives are used.

3. <u>Convenience Sampling</u>—This method is adopted when a 'convenient' group of individuals is used as the sample. For example, to select a sample of 40 students from the university, we can take the students from one class, or we can interview the first 40 students that we come across. This method of sampling is very biased and often leads to non-representative samples.

Data Collection:

Generally, data may be collected through one of the following three sources:

1. Published Sources: There are many published sources of data that we have access to. These may be available to us in printed books, monographs, and most importantly the Internet. Government agencies, international organizations, and many institutes provide publicly available sources on the Internet and there is a vast body of literature on these statistics.
2. Designed Experiment: This type of data collection is used when the researcher wants to examine the effect of a certain conditioning. Therefore in designed experiments, the researcher exerts strict control over experimental units.
3. Observational Study: This method is used when the goal of the study is to observe experimental units without control. Survey is an example of this type of data collection, where usually a questionnaire is designed and respondents answer the questions.

Exercise 1

A company has ordered a consignment of 10,000 identical digital components from abroad. When the shipment arrives, because of the time constraint, the quality control inspector selects only 200 components for inspection. On the basis of the number of defective items found in the selected items, the quality control inspector will either accept or reject the whole consignment.

a. This is an example of _____statistics.

Explain:

b. The 10,000 components form _____.

Explain:

c. The 200 selected components form _____.

Explain:

d. The method of sampling most appropriate for the inspector is _____.

Explain:

Section 3: Types of Data

Once the process of data collection is ended and all the information is gathered, two types of data are encountered:

> **Qualitative**: Each observation can be classified as being in one of several possible categories. Examples of this type of data are make of a car, ethnic background of an individual, and the blood type of a person. Qualitative data are also referred to as '**categorical**.'

> **Quantitative:** Each observation measures the numerical value of a variable of interest to the investigator. Examples of such a type of data include number of children in a household, the amount of fluid in a soda pop, number of checks that you write in a month, and water consumption of a household. This type of data is often referred to as '**numerical**' and can be broken down into two types:

1. <u>Discrete</u>: If the possible values of the variable of interest can be listed (e.g., integers) then we have a discrete set of data. In the above examples, number of children per household and the number of checks that you write in a month are examples of discrete numerical.

2. <u>Continuous</u>: If, on the other hand, the possible values can be any real number within a feasible interval then we have a continuous data set. In the above examples, the amount of fluid in a soda pop and water consumption of a household are examples of continuous numerical data. Other examples are time to perform a task, the height of a person, the weight of a bag of potato, and so on.

YOUR TURN! Please complete the required exercises below directly in this book. You are encouraged to discuss each exercise with a partner or group. You may be asked to complete these exercises in class or outside of class.

Exercise 2

Classify each variable as qualitative or quantitative:

1. Number of students in a class
   a. Qualitative   b. Quantitative

Explain:

2. Time that it takes to run a 100-meter dash
   a. Qualitative   b. Quantitative

Explain:

3. Brand of the shoe worn by your favorite sports player
   a. Qualitative   b. Quantitative

Explain:

4. Type of housing available to students in a college
   a. Qualitative   b. Quantitative

Explain:

Exercise 3

For the following quantitative variables, determine if they are discrete or continuous:

1. Number of text messages that you send in one day
   a. Discrete   b. Continuous

Explain:

2. Lifetime of a brand of dishwasher
   a. Discrete   b. Continuous

Explain:

3. Number of cookies in a one pound cookie jar
   a. Discrete   b. Continuous

Explain:

4. The exact weight of a 5-pound bag of potato
   a. Discrete   b. Continuous

Explain:

Section 4: Use of Statistical Software

Like all other disciplines, the advent of computers has had a profound impact in statistics. The use of computers has facilitated analysis of many large data sets that would have been impossible before. In the last 25 years several new powerful statistical methodologies have been introduced that are very computer intensive. A myriad of computer software have become available that have made advanced data analysis much more amenable. Among the most popular statistical software we can name:

MS EXCEL: Most common, but has limited statistical functionality. It is easy to use owing to its spread sheet format. Since this software is not specifically developed for statistical applications, it does not have the capability for performing many of the statistical techniques.

MINITAB: Has the spread sheet format and menus are geared toward statistical analysis. It is very simple to use. Minitab is most often used for educational purposes.

SPSS (Statistical Package for Social Sciences): Has many advanced features. It is most popular among social scientists and researchers in the field of education

SAS (Statistical Analysis System): Contains the most comprehensive state-of-the-art statistical procedures. It is used by most research centers, government agencies, and private organizations around the world. It requires programming skills.

R: A comprehensive object-oriented software. It has become very popular in recent years. It is available on the Internet as a free download.

Descriptive Statistics

Section 1: Introduction

As discussed in Chapter 1, **descriptive statistics** is the branch of statistics that deals with summarizing, organizing and hence describing the data set. Large data sets cannot meaningfully be utilized until they are summarized and organized in a presentable way. In **descriptive statistics**, we use many different types of charts, graphs, and numerical measures to make the data set more informative. These charts and graphs are popular methods of data representation. If you open any copy of a national or your local newspaper and go to the business section, you will more than likely see statistical charts and graphs are used to display business trends, market behavior, etc. Similarly, if you browse through the sports section, you will see that descriptive methods and measures are used to depict various aspects of sports statistics. They often discuss statistics of players, games, and much more. In this chapter we present some of the most common descriptive statistical methods. Recall that in the last chapter, we also discussed two general types of data—**Categorical** and **Numerical**.

Section 2: Categorical Data

To summarize categorical data, we set up a table we call a **frequency distribution** that shows the frequency of occurrence of each category in the data. To set up the frequency distribution, we list the categories and use tally marks or some other counting method to find the number of times each category has occurred in the data set.

■ Example 1: The Weather This Month—Table

During a 30 day period, the type of weather was categorized each day as Sunny (S), Cloudy (C), Rainy (R), or Stormy (T). Below you will see a list of the observed weather for 30 days.

$$S \ S \ C \ C \ R \ R \ C \ S \ S \ S \ S \ C \ T \ T \ R \ R \ C \ C \ S \ S \ S \ S \ S \ R \ R \ R \ C \ S$$

As you can see, the data gathered is not organized and answering questions about this month's weather could be difficult. This is exactly why we create a **frequency distribution.**

Category		Frequency	Relative Frequency
Sunny	┼┼┼ ┼┼┼ ││││	13	0.433
Cloudy	┼┼┼ │││	8	0.267
Rainy	┼┼┼ ││	7	0.233
Stormy	││	2	0.067
Total		30	1.00

How does this table help us? When we collect and organize data it is usually for the purpose of answering some questions. For example, we might be interested to answer questions like, "How many days was the weather sunny?" or "What percentage of the month was it raining?" We can look beyond data and ask questions like "Is this typical weather for this month?" We are able to use the table to answer the above questions as oppose to using the raw data.

It is important to note that the sum of all frequencies is 30 as it should be. If each datum (data point) is counted once and only once then the sum of all frequencies will be precisely equal to the total number of data in our set. The last column in the table gives the **relative frequency** of each category, which is the proportion of occurrences of each category. The **relative frequency** of each class is found by dividing the frequency of that class by the total number of measurements in the data set.

There are generally two common methods of graphical representation for categorical data; those are **Bar Graphs** and **Pie Charts.**

◼ Example 2: The Weather This Month—Bar Graph

Bar Graph: A bar graph can be constructed using the **frequencies** or the **relative frequencies**.
To draw a bar graph:

- ◼ We first draw horizontal and vertical axes.
- ◼ On the horizontal axis, we mark the *categories* of data. Note that since there is no scale on the horizontal line, the order in which the categories are presented does not matter.
- ◼ On the vertical axis, we put a proper scale based on the *frequencies* (relative frequencies). For each category, we then draw a vertical bar representing the frequency (relative frequency) of that category.

If the bar graph is presented in such a way that the frequencies (relative frequencies) are arranged in decreasing order, as in the bar graph below, we call that the Pareto diagram. The goal of the Pareto diagram is to display the most important (more frequent) categories first. The graphs were created using https://plot.ly/. Notice how we can once again answer questions about the weather as we did with the frequency distribution.

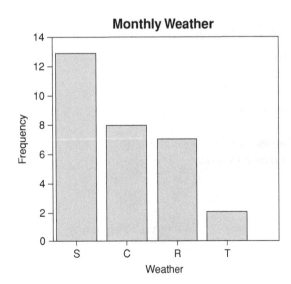
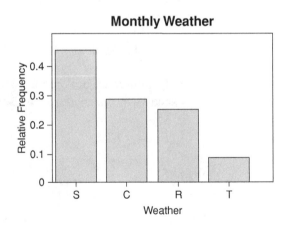

Pie Chart: Pie chart or pie graph is another way to represent the data pictorially. A pie chart can also be drawn using either the frequencies or the relative frequencies. In a pie chart, the area of a circle is used and relative frequencies are shown as pieces of a pie within the area of the circle. Figure 3 displays the pie chart for the weather data in relative frequency.

■ Example 3: The Weather This Month—Pie Chart

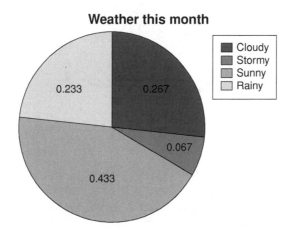

Weather this month

Legend:
- Cloudy
- Stormy
- Sunny
- Rainy

In practice, pie chart is a very often used and is a popular method of graphical representation. For example, in business, we can use a pie chart to show market shares or sales of different sectors of a company.

Critical Thinking: Can you think of an application of pie chart in sports?

Exercise 1

The attendant of a parking lot at a popular movie theater records the make of the first 45 cars that enter the lot:

Ford, Chevrolet, Ford, Chrysler, Toyota, Honda, Ford, Honda, Chevrolet, Toyota, Honda, Ford, Hyundai, Ford, Toyota, Honda, Chrysler, Toyota, Toyota, Ford, Chrysler, Ford, Nissan, Toyota, Nissan, Ford, Chevrolet, Ford, Honda, Ford, Honda, Ford, Ford, Toyota, Ford, Honda, Ford, Chrysler, Chevrolet, Chrysler, Ford, Ford, Toyota, Ford, Toyota

Set up a frequency chart for this data set by completing the table below.

Category	Frequency	Relative Frequency
Ford		
Chevrolet		
Chrysler		
Toyota		
Honda		
Hyundai		
Nissan		
Total		

Exercise 2

Using the frequency distribution for the make of the 45 cars in Exercise 1, draw

a. A bar chart using the frequencies

b. A pie chart using the relative frequencies

Ford   Chevrolet   Chrysler   Toyota   Honda   Nissan   Hyundai
Make of cars

Exercise 3

Answer the following questions with complete sentences. **Justify** your solution using either exercise1 or 2 or both.

a. How many cars was either Honda or Toyota? How do you know?

b. Are there more Chryslers or Chevrolets? How do you know?

c. What percentage of cars on the lot were Ford, Chrysler, and Chevrolet?

d. Foreign cars are considered to be from Honda, Toyota, Hyundai, or Nissan. Do you think more people buy foreign cars versus American cars? Why or why not?

Critical Thinking: Suppose you show a bar graph using the relative frequency of the make of cars to a friend. They ask, "How many cars were in the parking lot?" Can you answer this question using ONLY the relative frequency bar chart? Why or why not?

Exercise 4

Table below shows the frequencies of selected circulatory diseases among persons 18 years of age and over by age in the United States in 2005.
(*Source: Summary Health Statistics for US adults : National Health Interview Survey 2005. Center for Disease Control and Prevention, US Department of Health services, Series 10. Number 232, Page 17.*)

Age	# people	Heart Disease All Types	Heart Disease Coronary	Hypertension	Stroke
18–44 years	110,431	4,763	1,089	8,067	401
45–64 years	72,296	9,822	5,316	22,521	1,558
65–74 years	18,446	4,927	3,480	9,127	1,144
75 years and over	16,600	6,071	4,203	9,044	2,063

Below is the frequency distribution for Hypertension. There are missing values and your job is to fill in those missing values. Show work for each calculation.

Age	Frequency	Relative Frequency
18–44	8,067	0.165
45–64		
65–74	9,127	
75 and older		0.185
Total	48759	1

Using the original table answer the following questions.

a. How many people participated in this survey? Show work and explain.

b. What percentage of people surveyed have had a stroke? Show work and explain.

c. What percentage of 18–44 year olds surveyed have heart disease? Show work and explain.

Section 3: Numerical Data

As we described in Chapter 1, in a numerical data set each data point is the quantitative measure of a certain variable. Here, we discuss three methods used for graphical representation of numerical data.

Dot Plot

A **dot plot** is a simple but powerful way to present numerical data. Here is how you draw a dot plot:

✔ We draw a scaled horizontal line.
✔ We then represent each observation with a dot above the line.
✔ If a measurement is repeated more than once in the data set, the dots are placed on top of each other.

■ Example 4: The Number of Children per Household—Dot Plot

A survey is done on 25 randomly selected homes with the following question: "How many children under the age of 12 reside here?" The data was collected and the here are the results:

3	2	2	4	3	4	0	1	3	3
0	3	2	2	1	1	5	0	3	2
3	1	2	3	3					

As you can see each house gave a number and that number represents how many children reside in the house. The type of data you see in this example is called **discrete numerical** data, i.e., values or observations that is counted as **distinct and separate**.

Much like categorical data, there are ways to represent numerical data graphically. We will discuss how to display (discrete) numerical data in the following section.

As you can see in the dot plot to the right, there were 2 houses that had 4 children. A dot plot is useful in displaying the general pattern of the data.

Critical Thinking: Answer these questions.

○ *What does EACH dot represent? Write a complete sentence.*
○ *How many households had at LEAST 3 children per household? How do you know?*
○ *How many households had at MOST 3 children per household? How do you know?*

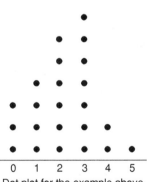

Dot plot for the example above

Stem-and-Leaf Display

To draw a stem-and-leaf display, we first split each datum (data point) into two parts; Stem and Leaf. We then draw a vertical line listing all stem values to the left of the line and leaf values to the right of the line.

■ Example 5: The Weight of Newborn Babies—Stem-and-Leaf Display

The weight at birth for 25 infants at a local hospital was recorded (in pounds) and rounded to the nearest ytenth of an ounce:

6.7	7.3	7.1	8.2	5.8	7.5	7.2	6.6	8.3	9.1
7.3	6.7	6.5	7.8	7.2	12.3	7.0	7.3	8.8	7.9
6.2	6.8	5.7	7.2	7.2					

Since the weights were rounded, the data set is again a discrete data set. Another way of displaying discrete data is a stem-and-leaf displaying.

The check marks below give an outline of how the stem-and-leaf display was constructed below.

✔ We noticed that the weights of the babies ranged from 5.7 to 12.3. To display properly, we must decide which are the stems and which are the leafs. As you can see, we split our data at the decimal place.

✔ The stem parts are then the integers between 5 and 12 and the leaf parts are integers between 0 and 9.

✔ After listing the stem values on the left side of the vertical line, each data point is represented once and only once by its leaf value on the right side of the line.

✔ We note that the units for stem and leaf are carefully stated in the display. **It is important to give the units for the stem values and the leaf values for otherwise it would not be possible for the reader to reconstruct the original data.**

Stem	Leaf
5	7 8
6	2 5 6 7 7 8
7	0 1 2 2 2 2 3 3 3 5 8 9
8	2 3 8 Stem: Ones
9	1 Leaf: Tenths
10	
11	
12	3

NOTE: It is important to note that larger sets of data may not be appropriately displayed by a stem and leaf. Also, when deciding on stem values, we have to select them in such a way that we do not have too few or too many stems since otherwise the stem-and-leaf display become uninformative. Although there is no rule, a stem-and-leaf display with between 5 and 15 stem values is considered to be reasonable.

Critical Thinking: Answer the following.

○ What does each STEM represent? What does each LEAF represent? Explain.

○ Determine how many babies were born LESS THAN 9 pounds? Explain.

○ As you look stem-and-leaf display, do you consider any baby to have an unusual weight? Explain.

You probably noticed that there was one baby whose weight was 12.3 pounds. This baby's weight seems rather unusual and not quite in tune with other values in the set. An **outlier** is a value in the data set, which is either significantly lower or significantly higher than all other values. Identification of outliers in a data set is an important step of data analysis since in most cases they should be scrutinized in order to find out about the cause for the occurrence of the outlier.

In Figure 4, we note that even though there were no data points with the stem values of 10 or 11, we listed these values to specifically show the gap and occurrence of an outlier. We will discuss a method of deleting outliers later in this chapter.

There are other applications of a stem-and-leaf display. One such application is in the comparison of two data sets. A 'comparative' stem-and-leaf display is often used for this purpose. To draw a comparative stem-and-leaf display to compare two data sets, we draw a common set of stem values. We then represent one set of stem values on the left and the other set of stem values on the right of the stem values. A comparative stem-and-leaf display is also referred to as a side-by-side stem-and-leaf display.

Outliers

■ Example 6: The Blood Pressure of Women versus Men—Side-by-Side Comparative Stem-and-Leaf Display

The systolic blood pressure (SBP) was measured for 25 men and 25 women, both groups being over 50 years of age is listed below. What follows is a side-by-side comparative stem-and-leaf display.

Men									Women							
137	126	135	121	155	124	144	131	147	135	153	133	164	166	174	163	152	121
118	182	133	135	142	144	139	137		133	138	142	155	163	171	177	162	
133	136	139	134	145	148	115	122		144	142	145	161	142	144	142	155	

There are several interesting features that are worth noting. By a visual comparison, we see that men overall seem to have lower SBP values, although there was one person who had a high of 182. While the distribution of the SBP for men seems to be more concentrated, the SBP distribution for women is more spread out with higher variability. We will discuss how to measure variability in a data set later.

Note also that the units for both the stem parts and the leaf parts are stated in the display. *Critical thinking: Answer the following:*

Why do you think the stem plots for the blood pressure for men versus women are different? What factors may cause this to happen?

Men	Stem	Women
5 8	11	
6 1 4 2	12	1
7 5 1 3 5 9 7 3 6 9 4	13	8 3 3 5
4 2 4 8 7 5	14	2 5 2 4 2 2 4
5	15	5 5 2 3
	16	1 2 3 3 6 4
	17	7 1 4
2	18	

Stem: Tens and Hundreds
Leaf: Ones

Exercise 5

Suppose that you work in a financial institution and are asked to recommend an amount to be deposited daily in the ATM machine. In order to make an informed decision for a reasonable amount, you obtain information on the total daily cash withdrawals in the last 30 days (in $100s)

58	64	76	87	46	89	92	81	73	77
55	85	74	68	59	61	92	88	88	73
71	69	83	84	89	73	66	134	75	79

a. What type of graphical representation is appropriate for this data? Explain briefly then provide that graphical representation. Be sure to include a title!

b. Comment on how the bank might use this graph to make an informed decision about the amount of cash that needs to be deposited in the ATM machine on a daily basis.

Histogram

A **histogram** is drawn based on the *frequency distribution* of the data set. As described before in the case of qualitative data, a frequency distribution provides the frequency of occurrence of each class or class interval of the data set.

Again, for the case of numerical data, we make sure that the number of classes is not too few or too many to make the frequency distribution non-informative. Similar to what we said about the frequency distribution of qualitative data, there is no set rule for the number of classes or class intervals, but once again, a number between 5 and 15 is considered reasonable.

■ Example 7: The Number of Children per Household—Histogram

Class		Frequency	Relative Frequency
0	|||	3	0.12
1	||||	4	0.16
2	⊬⊬ ||	6	0.24
3	⊬⊬ ||||	9	0.36
4	||	2	0.08
5			1	0.04
Total		25	1.00

Below you will see a table that displays the data from our previous example in which we asked the question, "How many children reside here?" In this example, we will describe how we will display this data using a **histogram**.

In our example regarding the number of children per household, we can regard the numbers 0, 1, 2, 3, 4, 5 as the classes of our data. Clearly, when we have numerical discrete data and if the number of possible values for the variable is a reasonable number, e.g., between 5 and 15, we can determine these classes very easily. However, if the possible values of the variable are a large set, or our data set consists of continuous data, as in our example about the birth weight, then the choice of classes is not trivial. In these cases, we divide a convenient range that contains all the data points into several subintervals. We choose these subintervals of equal length (unless doing so will make the histogram meaningless). The frequency of each class interval is then obtained in the usual manner. Once the frequency distribution is constructed, a histogram can be drawn. A histogram is defined as a connected sequence of rectangles drawn in such a way that the base of each rectangle represents a class or a class interval of the data and the height of the rectangle represents the corresponding frequency or relative frequency.

Below you will see a histogram of the number of children living at home.

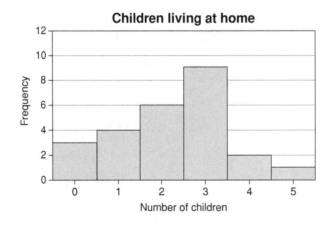

■ Example 8: Birth Weights—Histogram

In the birth weight data, the range 5 to 12.5 encapsulates the entire the data set. We therefore consider the class intervals 5.0 to 6.5, 6.5 to 8.0, 8.0 to 9.5, 9.5 to 11.0, and 11.0 to 12.5. Be mindful of the end points of your interval. Notice in the table below, there are two babies with birth weights from 5.0 to 6.5. If you go back and look at the raw data, you will see there is a baby that weighs 6.5 pounds. We have simply moved that weight to the second interval, 6.5 to 8.0. To be consistent, we will do the same for any other weights that are also an endpoint.

Class Intervals		Frequency	Relative Frequency
5.0–6.5	|||	3	0.08
6.5–8.0	|||| |||| |||| ||	17	0.60
8.0–9.5	||||	4	0.28
9.5–11		0	0.00
11.0–12.5			1	0.04
Total		25	1.00

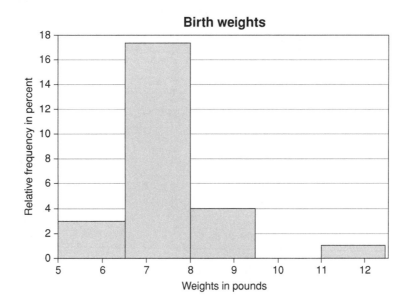

After we draw the histogram, what kind of shapes can we expect? First, we classify histogram shapes according to the number of peaks of a histogram. Each peak of a histogram is called a 'mode' of the distribution. If a histogram has one mode we call it unimodal, if it has two modes, we call it bimodal and if it has more than two modes we call it multimodal.

Critical thinking: Look at the histograms to the right and draw with your writing utensil a curved line the follows the top of the bars.

○ *How many modes are there for each histogram?*

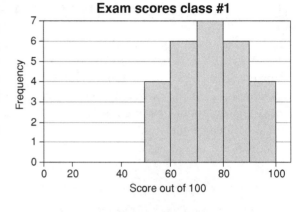

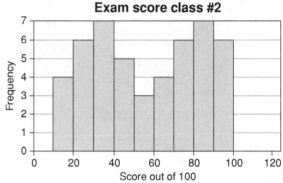

○ *What factors play a role in how exam scores are distributed?*

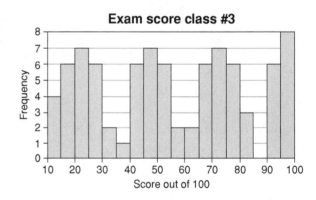

Most distributions that we encounter in practice are unimodal and there is a further classification of this group of distributions. A unimodal distribution is either symmetric or skewed. It is symmetric when one half is the image of the other half (see first picture below). If it is skewed, we call it **right (positively)** skewed if the long tail of the distribution is on the right and heavy tail is on the left (see middle picture below) On the other hand, we call the histogram **left (negatively)** skewed if the long tail is on the left and the heavy tail is on the right (see last picture below). It is also worth noting that the existence of outliers on the upper end of the distribution will make it right skewed and outliers in the lower end of the distribution will make it left skewed.

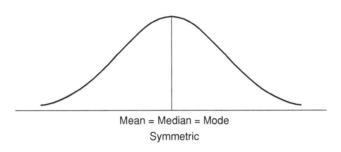

Mean = Median = Mode
Symmetric

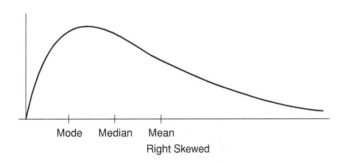

Mode   Median   Mean
Right Skewed

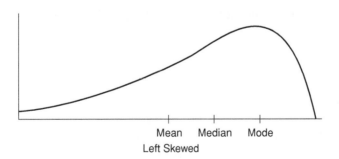

Mean   Median   Mode
Left Skewed

Exercise 6

Draw a relative frequency histogram for the ATM data using 55 to 70, 70 to 85, and so on as class intervals. Be sure to make a decision on what to do with any values that are end points.

58	64	76	87	46	89	92	81	73	77
55	85	74	68	59	61	92	88	88	73
71	69	83	84	89	73	66	134	75	79

Complete a–c

a. Describe the shape of the histogram.

b. What is the relationship between the median and the mean? How do you know?

c. What amount should be deposited daily in the ATM machine? Answer in complete sentences with a justification.

Section 4: Numerical Summary

In the last section we discussed different methods of pictorial summarization of the data. These methods provide the tools for visual examination of the information. Here, we consider numerical summarization, in particular we seek some numerical values that summarize the information contained in the data. First, we consider **numerical measures of center**.

How can we squeeze the information contained in the data set in one number? How can we determine a number that could represent a typical value of our dataset? One way is to measure the center of the data set. There are three methods of measuring the center of a data set: **mean, median** and **mode**.

The **mean,** denoted by $\bar{x}$, of a data set is the arithmetic average of the values in the data set. We use this measure all the time in our everyday life. To find the mean of a sample consisting of n data points, we first add all the numbers and divide the sum by n. Therefore if we call the first number in the set x_1, the second number x_2 and so on, all the way to the nth number x_n, then to find the mean we first add these numbers, $x_1 + x_2 + \cdots + x_n$ and divide this total by n. We use a symbol $\bar{x}$ to show the sample mean.

$$\bar{x} = \frac{(x_1 + x_2 + \cdots + x_n)}{n} = \frac{1}{n}(x_1 + x_2 + \cdots + x_n)$$

As you can see, you may rewrite this equation such that the $\frac{1}{n}$ is factored and you are left with the sum of the data values. In order to express the *formula* for the mean, we use a mathematical notation for summation. In mathematics, to denote the sum of variables, capital Greek letter sigma ($\sum$) is used. We can use this notation to describe the **sum** of the data values as

$$x_1 + x_2 + \cdots + x_n = \sum_{i=1}^{n} x_i$$

which means the sum of all $x_i's$ when the suffix i runs from 1 to n. Since n represents number of data points each x_i represents a single date point. Now we can replace the sum in our equation above:

$$\bar{x} = \frac{(x_1 + x_2 + \ldots + x_n)}{n} = \frac{1}{n}(x_1 + x_2 + \cdots + x_n) = \frac{1}{n}\sum_{i=1}^{n} x_i$$

We now have a formula (as opposed to just an equation) for the mean of a set of numbers.

$$\bar{x} = \frac{1}{n}\sum_{i=1}^{n} x_i$$

Often, when there is no confusion, we drop the suffixes and simply write

$$\bar{x} = \frac{1}{n}\sum_{i=1}^{n} x$$

which reads as the mean being equal to the sum of xs divided by n.

■ Example 9: The Time It Takes a Child to Read a Passage—Mean

A passage was given to 8 randomly selected first grade students and the time taken for each child to read the passage was recorded (in minutes):

12.6, 14.5, 15.1, 8.3, 11.9, 10.3, 9.9, 10.6

QUESTION: What is the mean of the data?
SOLUTION: To clarify the notation above, we can label the data using the x_i:

$$x_1 = 12.6, x_2 = 14.5, x_3 = 15.1, x_4 = 8.3, x_5 = 11.9, x_6 = 10.3, x_7 = 9.9, x_8 = 10.6$$

For these data, the sample mean is

$$\bar{x} = \frac{(12.6+14.5+15.1+8.3+11.9+10.3+9.9+10.6)}{8} = \frac{93.2}{8} = 11.65 \text{ minutes}$$

Critical thinking:

○ Does it matter which data point we label x_1? x_2? Discuss.
○ You notice that the mean you found is not actually a data point. Why is it okay that the mean is not part of the set? Discuss.

The **median,** denoted $\tilde{X}$ or M, is another method of measuring the center of a data set and it is defined as a **number** that divides the data set into two haves such that one half of the values in the data set are numerically lower than this median value and the other half are numerically higher. To find the median complete the following steps:

✔ Order the date from smallest to largest.
✔ If the number of measurements is an odd number, we pick the middle value to be the median.
✔ If the number of measurements if an even number, we take the two middle values and find the average of those two numbers and that number is the median.

■ Example 10: The Time It Takes a Child to Read a Passage—Median

To find the median reading times

$$8.3, 9.9, 10.3, \textbf{10.6, 11.9,} 12.6, 14.5, 15.1$$

Because there are 8 measurements, the median is the average of 10.6 and 11.9 and thus $\tilde{x} = 11.25$ minutes.

Critical thinking: You notice that the median you found is not actually a data point. Why is it okay that the median is not part of the set?

Comparing Mean and Median

Both mean and median are very popular and useful measures of the center. Question that arises now is what the relative advantages and disadvantages of each of the two measures of the center are and which measure is preferred in what specific setting. **Here we provide a brief comparison of the two measures**. Generally, mean is considered to be a more informative measure since in the computation of the mean we use every single measurement while after ordering, the median is based on only one or two numbers in the data set.

In our example about the reading times of first graders, median would not change if say 15.1 is changed to any number equal to or higher than 14.5, but the value of mean would certainly change. This implies that the mean is more **sensitive to changes** in the data set. Specifically, the mean is very sensitive to **outliers**. Recall that outliers are numbers that are not quite in tune with all other numbers in the set. To illustrate, let us consider the following example.

■ Example 11: Salaries of Employees

The annual salaries of 5 randomly selected employees of a company (in $1000s) were as follows:

$$43.1, 38.5, 41.2, 88.6, 44.5$$

For these numbers the mean is the sum of the numbers that is 254.9 divided by 5 which gives us 51.18 thousand dollars, whereas the median is found by first ordering the salaries as

$$38.5, 41.2, 43.1, 44.5, 88.6$$

and picking the middle position which is 43.1 thousand dollars.

In the above example, which of the two numbers better represent the salary distribution in the company? Clearly in this case the median provides a better picture of the salaries. The value of the mean is inflated because of the existence of the outlier 88.6 in the data and is misleading. In general, we can use both mean and median as measures of the center although mean is more popular. However, if there are any known outliers in the data, then median would probably be preferred. It is further noted that in a completely symmetric distribution, the mean and median are the same. If the distribution is right skewed, then because of the existence of outliers in the upper end of the distribution, the value of the mean is higher than the median (as in the above example about the salary distribution in a company). If, on the other hand, the distribution in left skewed, the outliers are in the lower end and the mean is pulled towards the outlier and therefore the median would be a higher number than the mean.

Exercise 7

What is the mean? Show every calculation and step.

What is the median? Show every calculation and step.

Critical Thinking: Which of the two numbers better represent the salary distribution in the company?

In general, we can use both mean and median as measures of the center although mean is more popular. However, if there are any known outliers in the data, then median would probably be preferred. It is further noted that in a completely symmetric distribution, the mean and median are the same. If the distribution is right skewed, then because of the existence of outliers in the upper end of the distribution, the value of the mean is higher than the median (as in the above example about the salary distribution in a company). If, on the other hand, the distribution in left skewed, the outliers are in the lower end and the mean is pulled toward the outlier and therefore the median would be a higher number than the mean.

Exercise 8

In Exercise 5 you were asked to draw a relative frequency histogram for the ATM data. Use the histogram to complete a–c below:

a. Describe the shape of the histogram.

b. What is the relationship between the median and the mean? How do you know?

c. What amount should be deposited daily in the ATM machine? Answer in complete sentences with a justification.

Exercise 9

The number of cars that pass through an intersection was determined for 11 randomly selected days in a month:

$$64 \quad 28 \quad 47 \quad 87 \quad 55 \quad 69 \quad 76 \quad 35 \quad 162 \quad 71 \quad 54$$

a. Find the mean of the data using the formula: $\bar{x} = \frac{1}{n} \sum_{i=1}^{n} x_i$

Before you start: What is n? Why? What does each x_i mean?

b. Find the median:

Before you start: How many data points do you have? Why is finding the answer to this question important?

There is yet another method to measure the center of the data set, and that is by using the **mode**. The mode of a distribution is defined as **the most frequent value in the data set**. Therefore in the frequency distribution, the class with the highest frequency is the mode. Note that if the frequency distribution is based on class intervals as in continuous data, then the class interval with the highest frequency is the **modal class**. Thus in our example of number of children per household, 3 is the mode with the highest frequency of 9 and in our birth weight example, 6.5 to 8.0 is the modal class.

Measures of the center, specifically mean and median, give us information about 'typical' values in our data set. But, they do not provide any information about how close or how dispersed the values are in the set and as such they do not provide complete characterization of the data set. Here we give a motivational example for further discussion.

Measures of Dispersion and Variability

■ Example 12: Comparing Two Sets with Identical Mean and Median

Consider the following two hypothetical data sets:

Set I: 10, 20, 30, 40, 50   Set II: 28, 29, 30, 31, 32

For set I, the mean is, $\dfrac{10+20+30+40+50}{5} = 30$ and the median is the middle number which is also 30. Incidentally, mean and median are identical because the data set is completely symmetric.

For set II, we have, $\dfrac{28+29+30+31+32}{5} = 30$ and the median is again 30. But, by looking at the two sets, we see that they have completely different characteristics even though they have identical means and medians.

In one set, the numbers are much closer to each other compared to the other set. If I tell you that the average age in a neighborhood is 30, you have no information about the **variability** of ages in that neighborhood. You don't know if the neighborhood has ages similar to set I or set II. For this reason, in this section, we define **measures of variability**.

One way to differentiate between the two data sets above is to realize that in set I, the numbers start at 10 and go to 50 and in set II the numbers start at 28 and go to 32. We therefore define the **range** of a data set as the **difference between the largest and the smallest values**.

Range = Largest–Smallest

Using this definition we see that range for set I is 40 and the range of set II is 4.

We now see the difference between the two data sets. If I tell you that the average age in a neighborhood is 30, and the range is 40, you know that there are individuals with varying ages residing in the neighborhood. But, if I tell you that the average age in a neighborhood is 30 and the range is only 4, then you know that everyone in that neighborhood is around 30 years of age. Therefore the range can be an informative way of measuring variability.

There is a major problem with the range. Because in the calculation of range we only use the extreme values in the data set (the largest and the smallest), the range is **highly sensitive to outliers**. If there is an outlier in the data, the value of range will be significantly inflated and can be misleading. For this reason, the range is only used with caution.

When we want to measure **variability** in a data set, it makes sense to think of variation with respect to a frame of reference. Clearly the center of a data set is a sensible value to use as reference. Although we defined different methods of measuring the center of a data set, here we use the mean as the reference to measure variability. Let us denote the sample data with n measurements as $x_1, x_2, \ldots x_n$ and let $\overline{x}$ be the sample mean. We subtract the mean from each data point to measure how far that value is from the center, i.e., to measure its variation from the mean,

$$x_1 - \overline{x},\ x_2 - \overline{x}, \ldots, x_n - \overline{x}$$

We call these values the ***deviations***.

■ Example 13: The Time It Takes a Child to Read a Passage—Deviations

Consider the example we discussed earlier on the reading times of 8 first graders. Recall that we had

$$12.6, \ 14.5, \ 15.1, \ 8.3, \ 11.9, \ 10.3, \ 9.9, \ 10.6$$

where $\bar{x} = 11.65$ minutes. We can use Excel to calculate the deviation for each data point.

Data points (x_i)	Mean ($\bar{x}$)	Deviations ($x_i-\bar{x}$)
$x_1 = 12.6$	11.65	12.6–11.65 = 0.95
$x_2 = 14.5$	11.65	2.85
$x_3 = 15.1$	11.65	3.45
$x_4 = 8.3$	11.65	−3.35
$x_5 = 11.9$	11.65	0.25
$x_6 = 10.3$	11.65	−1.35
$x_7 = 9.9$	11.65	−1.75
$x_8 = 10.6$	11.65	−1.05

We might think that perhaps the **average of these deviations** can provide a measure of variability. But, if we add the deviations in the above example we get 0. Is that coincidental? Well, no. The reason is that some of the deviations are positive and some are negative. Because the mean is the center of the data, the positive and negative values offset each other and we get 0. **In fact,** *for any data set the sum of deviations is 0*, that is using our sigma notation, we can write

$$\sum (x - \bar{x}) = 0$$

Therefore if we can somehow make all these values positive, we will resolve the problem. **There are two approaches that we can take**. One way is to consider the absolute value of the deviations. We know that the absolute value of a positive number is positive and the absolute value of a negative number is also positive. Another approach is to take the square of each deviation since squaring a positive or a negative number will produce a positive number. Here, we take the latter approach. We square the deviations and average them and use the average as a measure of variability. We give this average a name and call it the ***sample variance***. In averaging the square deviations, since there are n items being averaged, naturally we would expect to divide the sum of square of deviations by n. However, theoretically it can be proved that dividing the sum of square deviations by n would underestimate variability and a more unbiased estimate of variation is derived if instead we divide by $n-1$. Therefore we define the sample variance as

$$s^2 = \frac{(x_1 - \bar{x})^2 + (x_2 - \bar{x})^2 + \cdots + (x_n - \bar{x})^2}{(n-1)} = \frac{\sum (x - \bar{x})^2}{(n-1)}$$

Note that we use the symbol s^2 to show the variance in order to emphasize the fact that the deviations are squared and that ***variance is always a positive quantity***. Because of this squaring, the unit of measurement of variance is not the same as the original data and variance is measured by units that are the square of the original data. Therefore if our sample data are measured in say minutes (as in time to read a passage) then variance is measured in terms of (minutes)^2. Similarly if the data are in measured in say pounds (as in birth weight) the variance is in (pounds)^2. Therefore the interpretation can become rather difficult. In order to have a measure of variability that has the same units of measurement as the original data, we take the square root of variance and use that as a measure of dispersion. We give that a name and call it the '**standard deviation**' of the data. Therefore we define the **standard deviation** as

$$s = \sqrt{s^2} = \sqrt{\frac{\sum (x - \overline{x})^2}{n - 1}}$$

■ Example 14: The Time It Takes a Child to Read a Passage—Variance and Standard Deviation

For illustration, we compute the variance and standard deviation of the measurements of the time that it takes first graders to read a passage. We recall that the deviations are

$$0.95,\ 2.85,\ 3.45,\ -3.35,\ 0.25,\ -1.35,\ -1.75,\ -1.05$$

Therefore squaring each of these numbers, adding the squares and dividing the total 7 find the variance,

$$S^2 = \frac{(.95)^2 + (2.85)^2 + (3.45)^2 + (-3.35)^2 + (0.25)^2 + (-1.35)^2 + (-1.75)^2 + (-1.05)^2}{8 - 1} = 5.46 \text{ (minutes)}^2$$

and the standard deviation is given by

$$S = \sqrt{5.46} = 2.34 \text{ minutes}$$

Critical Thinking: In your own words, interpret 2.34 minutes in the context of the example.

Exercise 10

We once again examine the number of cars passing through an intersection for 11 days.

64 28 47 87 55 69 76 35 162 71 54

a. Find the range for this dataset—show all work.

b. Find the variance and standard deviation:

In the calculation of variance, arranging the dataset in a table as below is useful

x	$x - \overline{x}$	$(x - \overline{x})^2$
64		
28		
47		
87		
55		
69		
76		
35		
162		
71		
54		

Then, the variance is

$$s^2 = \frac{\sum (x - \overline{x})^2}{n - 1} = $$

and the standard deviation is

$$S = \sqrt{S^2} = $$

Computation of variance could be somewhat lengthy and laborious. We can reduce the amount of computation by using the **shortcut formula**, which is given by

$$s^2 = \frac{\sum x^2 - \frac{\left(\sum x\right)^2}{n}}{n-1}$$

Critical Thinking: Before we go on, take a minute to study the formula.
How is it the same?

How is it different?

The advantage of this shortcut formula is that we bypass the process of finding deviations and squaring them and thus shorten the computation time. Statistical software and calculators use the shortcut formula for the computation of variance as it is less intensive. In this formula we need the sum of the square of each measurement in addition to the sum of all measurements. Algebraically we can prove that this formula and the formula given earlier for the variance are identical. However, the proof is outside the scope of this course. We will illustrate how they are identical using a prior example.
From before we have for the reading times

$$12.6, \ 14.5, \ 15.1, \ 8.3, \ 11.9, \ 10.3, \ 9.9, \ 10.6$$

Now,

$$\sum x^2 = 12.6^2 + 14.5^2 + 15.1^2 + 8.3^2 + 11.9^2 + 10.3^2 + 9.9^2 + 10.6^2 = 1123.98$$

and

$$\sum x = 12.6 + 14.5 + 15.1 + 8.3 + 11.9 + 10.3 + 9.9 + 10.6 = 93.2$$

Therefore

$$s^2 = \frac{1123.98 - \frac{93.2^2}{8}}{7} = 5.46 \quad (\text{minutes})^2$$

which, as expected, is the same as before.

Exercise 11

For the number of cars passing through the intersection data, compute the variance and the standard deviation using the shortcut formula. And verify that your answer is the same as what you found for the variance earlier. Show all work.

$$64 \quad 28 \quad 47 \quad 87 \quad 55 \quad 69 \quad 76 \quad 35 \quad 162 \quad 71 \quad 54$$

$$\sum x =$$

$$\sum x^2 =$$

$$s^2 = \frac{\sum x^2 - \frac{\left(\sum x\right)^2}{n}}{n-1} =$$

Using Standard Deviation to Learn About the Data

Now that we have learned how to compute the standard deviation, **the question is how we can use it in order to learn about the distribution of the data**? Here we introduce two methods of using the mean and standard deviation to interpret the data.

A. Chebyshev's Rule

Chebyshev, a Russian mathematician, was first to use the standard deviation for interpreting the data in the following way. Let $\bar{x}$ and s be respectively the mean and standard deviation of the sample. Chebyshev proved that the interval two standard deviation about the mean, i.e., $(\bar{x} - 2s, \ \bar{x} + 2s)$ contains at least three-fourth of all the measurements. That is to say that at least 75% or ¾ of data are within two standard deviations of the mean. Chebyshev also proved that the interval three standard deviations around the mean i.e. $(\bar{x} - 3s, \ \bar{x} + 3s)$ contain at least eight-ninths of the measurements. In other words, at least 89% or $\frac{8}{9}$ of data are within three standard deviations of the mean.

In fact Chebyshev proved a more general theorem. He showed that if k is any real number, then the proportion of data within k standard deviations of the mean, i.e., $(\bar{x} - ks, \ \bar{x} + ks)$ is at least $1 - \dfrac{1}{k^2}$. This result is quite interesting and useful since with information only about the mean and standard deviation, we can estimate the proportion of data within any desired number of standard deviations. Conversely, if a measurement is within a certain number of standard deviations from the mean, we can determine the likelihood of its occurrence.

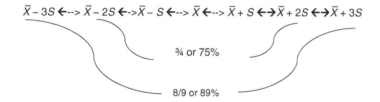

■ Example 15: Exam Scores

A professor announces in class that scores on a recent test had an **average of 72** with the **standard deviation of 8**. We are interested in answering the following questions:

a. What proportion of students scored between 56 and 88?
b. What proportion of students scored between 48 and 96?
c. What proportion of students scored lower than 48?
d. What proportion of students scored higher than 88?
e. What proportion of students scored higher than 84?
f. What proportion of students scored between 59 and 85?

SOLUTION: In order to understand how to answer these questions, we will go over how the solutions are calculated.

a. We first revisit the number line and use our number sense to answer part a.

    ✔ Using the number line below put a **square** around the average (mean) of the exam.
    ✔ Next, put a **circle** around the numbers 56 and 88.

✔ Now describe how the numbers 56, 72, and 88 are related.

✔ How can you incorporate the number "8" into your description above?

As we discovered on the previous page, the interval from 56 to 72 is of length 16, i.e., two standard deviations below the mean. The interval from 72 to 88 has a length of 16 also, i.e., two standard deviations above the mean. Therefore the interval (52, 88) can be written as $(\bar{x} - 2s, \bar{x} + 2s)$ where

$$\bar{x} - 2s = 72 - 2(8) = 52 \qquad\qquad \bar{x} + 2s = 72 + 2(8) = 88$$

To complete the answer for part a. we state: according to Chebyshev's theorem, **at least** 75% of the scores will fall in the interval (52, 88).

If you think about it, what you noticed on the number line can be described algebraically. We want to know "**how many standard deviations**" away a data point is from the mean. This can be found by the following:

$$\frac{(x - \bar{x})}{s}$$

That is, take the data value, **x**, and the mean, $\bar{x}$ and find the difference. Now you take the difference and divide it by the standard deviation.

If we look at the numbers for part and use $\bar{x} = 72$ and s = 8, we get

$$\frac{(56 - 72)}{8} = -2, \qquad\qquad \frac{(88 - 72)}{8} = +2$$

As you can see, the "2" means that the date value has a distance of 2 standard deviations from the mean. *Critical Thinking: The 2 is negative on one and positive on the other. What does this mean?*

b. For part b, we have the numbers 48 and 96

$$\frac{(48 - 72)}{8} = -3, \qquad\qquad \frac{(96 - 72)}{8} = +3$$

According to Chebyshev's theorem this implies that at **least 89%** or **at least** 8/9 of test scores were between 48 and 96.

c. Since **at least** 89% of test scores are between 48 and 96, this means that at **most 11%** are either less than 48 or higher than 96. As you can see on the number line below, the black line represents the interval from 48 to 96.

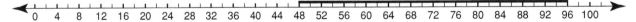

We may be tempted to think that perhaps about half of that amount, i.e., 5.5% of test scores are less than 48, but since we have no information about the shape of the distribution of the test scores, there is no reason to think that the underlying distribution is symmetric and hence it would be erroneous to associate 5.5% of test scores as being below 48 and 5.5% above 96. **All that we can say is that at most 11% of test scores are below 48.**

d. Similar to the argument that we made in part c. and using the result of part a., here we can say that **at most 25%** of test scores are above 88.

e. First, we need to find by how many standard deviations 84 is above or below the mean. We have

$$\frac{(84-72)}{8}=1.5$$

This results in 1.5 standard deviations away! This is slightly different than what we have experienced in parts a–d however, we still use Chebyshev's theorem. In this case, to find the proportion of test scores that are within 1.5 standard deviations of the mean, we substitute k = 1.5 in the **Chebyshev's formula**, to get

$$1-\frac{1}{k^2}=1-\frac{1}{1.5^2}=0.56$$

which means that **at least 56%** of test scores are within 1.5 standard deviations. Therefore, **at most 44%** of test scores are above 84.

f. We have

$$\frac{59-72}{8}=-1.625,$$

and

$$\frac{85-72}{8}=1.625$$

and therefore once again to find the proportion of test scores within 1.625 standard deviations of the mean, we substitute this number in the Chebyshev's formula,

$$1-\frac{1}{k^2}=1-\frac{1}{1.625^2}=0.621$$

Therefore **at least 62.1%** of test scores are between 59 and 85.

> **YOUR TURN!** Please complete the required exercise below directly in this book. You are encouraged to discuss each exercise with a partner or group. You may be asked to complete the exercise in class or outside of class.

Exercise 12

Based on a random sample of residents in a town, it was found that the average monthly expenditure on childcare per household was \$543 with a standard deviation of \$168, that is $\bar{X} = 543$ and $S = 168$. Based on this information, what percentage of households spend

a. Between \$207 and \$879?

By the Chebyshev's rule, _____%_____ of homes spend between \$207 and \$879 monthly on childcare.

b. In excess of \$879?

c. Between \$316.2 and \$769.8?

$k =$ _____

and by the Chebyshev's formula the proportion is given by

Hence the desired percentage is _____.

B. Empirical Rule

What makes the Chebyshev's theorem important and attractive is its generality. It does not assume anything about the shape of the distribution and can be applied to any kind of distribution, whether it is unimodal, bimodal, multimodal or if it is symmetric, right skewed, or left skewed.

Because of this generality, it is not very precise. Since in practice many of the distributions are (or can be well approximated) by symmetric unimodal mound shaped or what we call a **normal distribution**, we introduce another rule that only applies to this particular class of distributions, but is more precise than the Chebyshev's rule.

If the histogram of the data has (at least approximately) the shape of the **normal curve**, then

(i)   About 68% of the data are **within** one standard deviation of the mean.
(ii)   About 95% of the data are **within** two standard deviations of the mean.
(iii) Essentially all the data or approximately 99.7% are **within** three standard deviations of the mean.

The three listed items above are commonly referred to as the **Empirical Rule**.

(iv)   Since the mean and the median are equal, 50% of the data lie above the mean and 50% of the data lie below the mean.

The figure below demonstrates what a normal curve looks like according to the mean and standard deviation of the data.

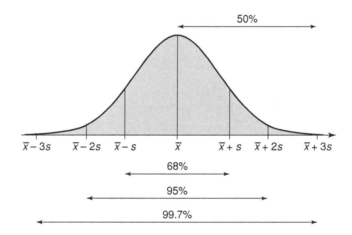

■ Example 16: Exam Scores—Continued

In the last example regarding the test scores, suppose that in addition to the mean and standard deviation of scores, the professor also announces that the histogram of the test scores closely resembled a normal curve. Then what can be said about the proportion of test scores that are:

a. Between 56 and 88?
b. Above 80?
c. Between 56 and 64?
d. Below 56?
e. Above 96?
f. Between 64 and 72?

❖ **SOLUTION**: Whenever you are asked to determine the proportions of data in a normal distribution, it is important to draw a normal curve and place the appropriate numbers.

By comparing the graph of the normal distribution for the test scores with the graph of the normal distribution that describes the Empirical rule, we can easily see that the answers to the parts of this problem are:

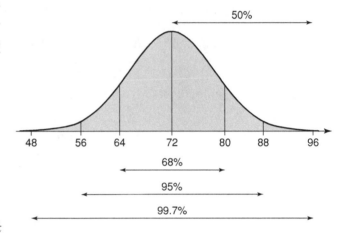

a. About 95
b. About 16%
c. About 13.5%
d. About 2.5%
e. Essentially none or approximately 0.15%
f. About 34%

Critical Thinking: Take a minute to pause and think about how the answers are from a to f. Use the curve provided. Write any additional notes in the space below.

Exercise 13

In the childcare monthly expenditure problem, assume that the distribution of monthly expenditures on child care for all families is a normal curve. Determine the percentage for each of the following. A normal curve has been provided for you to make notes on. Show all work for each answer.

a. More than $543
b. Between $207 and $879
c. More than $879
d. Less than $375
e. Higher than $1047

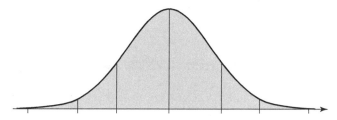

Numerical Measures of Relative Standing

In the scores on an exam example, of course, it would be useful to know what proportion of students scored within a given range. **But, it is more important to an individual student to know how his/her score compares with other scores in class**. In other words, we are often interested to know the *relative standing* of a specific measurement as compared with other measurements. Generally, there are two methods for measuring relative standing: Percentile Rank and Z-score.

A. Percentile Ranking

A measurement X is said to have a **percentile ranking** of P if P% of measurements in the data set are numerically lower than or equal to X and (100–X)% are numerically higher than X. Therefore if N is total number of data points and X is higher than or equal to n of them, then the percentile ranking of X is found by

$$P = \left(\frac{n}{N}\right) \cdot 100$$

Percentile ranking is often used for announcing results of standardized tests. For example, in high schools when students take a state test, the results are often announced in terms of percentile ranking. Specifically, if your percentile ranking in say math is 82, it means that your score was higher or equal to 82% and lower than 18% of all those who took the test.

■ Example 17: Myra's Score—Percentile Ranking

In an aptitude test, there were 344 applicants. Myra's score higher than or equal to 191 of them. Myra's percentile ranking is given by

$$P = \frac{191}{344} \cdot 100 = 55.5 \approx 56$$

Therefore Myra's percentile ranking is 56.

B. Z-Score

To find the **Z-score** of a measurement we first subtract the mean $\bar{x}$ from that measurement and divide the result by the standard deviation.

$$z = \frac{(x - \bar{x})}{s}$$

You may recall that when we discussed the Chebyshev's rule, we used this methodology to find the number of standard deviations by which a measurement is above or below the mean. Thus if the Z-score of a measurement is negative, it shows that the measurement is below the mean whereas if the Z-score is positive it indicates that the measurement is above the mean. The magnitude of the Z-score tells us how close or how far that particular measurement is away from the mean, *specifically by how many standard deviations.*

■ Example 18: Scores on an Exam—Z-score

Let us consider our test scores example again. We recall that the test scores had an average of 72 with a standard deviation of 8. Suppose April scored 83 on the test while her friend Jake received 54. We want to use Z-scores to interpret their scores.

For April
$$z = \frac{83-72}{8} = 1.375,$$

which shows that April's score is 1.375 standard deviations above the mean.

For Jake we have

$$z = \frac{54-72}{8} = -2.25,$$

which tells us that Jake's score was 2.25 standard deviations below the mean.

If we further know that the test scores had a normal distribution, then we can use the empirical rule and say even more about April's and Jake's scores. Since April's score is more than one standard deviation above the mean, her score is among the top 16% of the class. Similarly, since Jakes score is more than two standard deviations below the mean, then again by the empirical rule, his score is among the 2.5% of the class. In fact we can see that the empirical rule can actually be stated in terms of Z-scores.

Accordingly, we have the following restatement of the empirical rule (based on Z-scores):

If the distribution of data closely resembles the normal curve, then

(i)   About 68% of Z-scores are between −1 and +1.
(ii)   About 95% of Z-scores are between −2 and +2.
(iii) Essentially all or about 99.7% of Z-scores are between −3 and +3.

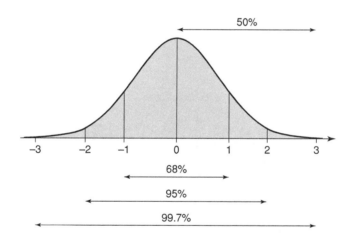

Exercise 14

All 173 graduating seniors took the SAT test in a high school. Their combined score in math and verbal had an average 1060 with a standard deviation of 140. Paul's score was 1210 and was exceeded only by 52 students.

a. Compute Paul's percentile ranking (show all work).

b. Compute Paul's z-score (draw a normal curve and show all work).

Quartiles, IQR and the Five-Number Summary

We learned how to use the mean and standard deviation to summarize the data. However, we know at the same time that mean, and to some extent the standard deviation, are somehow sensitive to outliers. We have seen that in the presence of outliers the mean could in fact be somewhat misleading. Here, we introduce a different method of summarizing the data set which not only is *insensitive to outliers*, but also identifies presence of outliers.

Recall that median was defined as a value determined in such a way that one half or 50% of the data are numerically higher and one half are numerically lower than that value. In a sense we can think of the median as a value that divides our data set into two equal halves. In a similar way we define the lower and upper quartiles.

Lower Quartile: A number determined in such a way that one quarter (25%) of data points are numerically lower and three quarters (75%) are numerically higher than that value is called the lower quartile. We use the symbol Q_1 to show the lower quartile (you may also see Q_L used).

Upper Quartile: A number determined in such a way that three quarters (75%) of data are numerically lower and one quarter (25%) are numerically higher than that value is called the upper quartile. We use the symbol Q_3 to show the upper quartile (you may also see Q_U used).

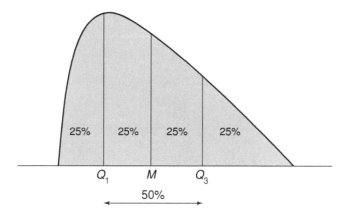

In order to determine the quartiles we complete the following steps:

✔ We first order the measurements from the lowest to the highest as we did for finding the median.
✔ We next identify the median position, which essentially gives us the lower half and the upper half of the data set.
✔ We then take the lower half and find its median, which is the lower quartile Q_1.
✔ Lastly, we take the upper half of the data set and find its median, which in turn gives us the upper quartile Q_3.

Note that in determining the median position to identify the lower and upper halves of the data set, if the number of data points in the data set is an odd number, then the median is uniquely defined as the middle value and the question is whether this median value should be counted as a data point in the lower half of the data set or the upper half. There is a controversy in this issue and some statisticians suggest that this middle value should not be counted either way while others suggest that the middle value should be included both toward the lower half and also toward the upper half of the data. Here, we adopt the latter and count the middle value both ways. Note that this problem does not occur if the number of data points is an even number since in that case once the median position is identified, we have an equal number of measurements below and above the median value.

Inter-quartile Range (IQR): We define the difference between the upper and lower quartiles as the inter-quartile range. It is the range of the middle 50% of all measurements and therefore it is a good representative of a data set. The advantage of the IQR is that it is not affected by outliers. In fact in many organizations this measure is used to represent to provide information about some characteristic of that organization. For example, to provide a profile of their students, many universities and colleges use IQR of the SAT scores of their incoming freshmen. Similarly, many companies use IQR to provide information about the salary range of the middle 50% of their employees.

■ Example 19: Commute to work—Quartiles and IQR

Distance travelled to and from work was determined for 15 randomly selected employees of a company (in miles):

$$22, \ 12, \ 31, \ 18, \ 8, \ 15, \ 11, \ 28, \ 2, \ 40, \ 18, \ 25, \ 15, \ 88, \ 10$$

For this dataset we wish to determine the quartiles and the IQR. First, if we order these numbers, we have

$$2, \ 8, \ 10, \ 11, \ 12, \ 15, \ 15, \ 18, \ 18, \ 22, \ 25, \ 28, \ 31, \ 40, \ 88$$

and because we have an odd number of values in the set, the median is the middle value, i.e., the 8th measurement that is 18. Now, as explained before, to have equal number of measurements in lower and upper halves of the dataset, we include the median both ways for finding the lower and upper quartiles. Hence the lower half consists of

$$2, \ 8, \ 10, \ 11, \ 12, \ 15, \ 15, \ 18$$

The median of which is the lower quartile. Therefore

$$Q_1 = \frac{11+12}{2} = 11.5 \text{ miles}$$

Similarly, the upper half of the dataset is

$$18, \ 18, \ 22, \ 25, \ 28, \ 31, \ 40, \ 88$$

And consequently the upper quartile is given by

$$Q_3 = \frac{25+28}{2} = 26.5 \text{ miles}$$

And finally, the IQR is

$$\text{IQR} = 26.5 - 11.5 = 15 \text{ miles}$$

This means that the company estimates that 50% of the employees of the company travel between 11.5 and 26.5 miles to work. About 25% travel less than 11.5 miles and about 25% travel more than 26.5 miles.

Five-Number Summary: The quartiles together with the median provide information about the middle 50% of the distribution. By adjoining the extreme values, i.e., the lowest and the highest values to them, we have the five-number summary that provides a good overall impression of the dataset. Therefore the five-number summary consists of

$$\textit{Min,} \quad Q_1, \quad \textit{Median,} \quad Q_3, \quad \textit{Max}$$

Boxplot and Identification of Outliers

A. Drawing a Boxplot

A very useful tool for pictorial representation of the *five number summary* is the boxplot. Boxplots can be drawn horizontally or vertically. Often boxplots are used for comparing data sets as well. Here we demonstrate the method for drawing a horizontal boxplot. The method for vertical boxplot is very similar. To draw a horizontal boxplot, we go through the following steps:

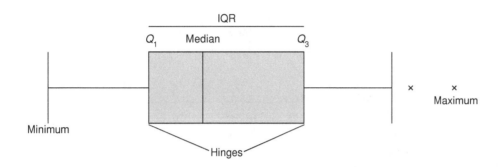

- ✔ Draw a scaled horizontal line
- ✔ Represent the lower and upper quartiles with a pair of vertical lines and form a box
- ✔ The lines representing the quartiles are called the **hinges** of the boxplot. From the mid-point of hinges, draw horizontal lines extending out 1.5 IQR on each side i.e. calculate the two quantities

$$Q_1 - 1.5\ IQR \qquad \text{and} \qquad Q_3 + 1.5\ IQR$$

  to know how far the lines should extend out. These lines are called the **whiskers** of the boxplot.
- ✔ The endpoints of whiskers are called the **inner fences** of the boxplot. From the inner fences, mark the points extending out another 1.5 IQR. Thus the points are marked at distances calculated from

$$Q_1 - 3.0\ IQR \qquad \text{and} \qquad Q_3 + 3.0\ IQR$$

These points are called the **outer fences** of the boxplot.

■ Example 20: Commute to Work—Five-number Summary

In our example regarding the number of miles commuted to work, the five-number summary is given by

$$2, \quad 11.5, \quad 18, \quad 26.5, \quad 88$$

With this information, the company has a rather clear picture of the distribution of the number of miles the employees travel to work. They know that about 25% of employees travel between 2 and 11.5 miles, about 25% travel between 11.5 and 18 miles and so on.

■ Example 21: Commute to Work—Box Plot

Once again, using the information about the commuting miles of the 15 employees of the company we create a box plot.

Minimum	2
Q_1	11.5
Median	18
Q_3	26.5
Maximum	88
IQR	15

Upper inner fence	$26.5 + (1.5 \times 15) = 49$
Lower inner fence	$11.5 - (1.5 \times 15) = -11$
Upper outer fence	$49 + (1.5 \times 15) = 71.5$
Lower outer fence	$-11 - (1.5 \times 15) = -33.5$

We should note that any **negative values** are not considered. Why is this? Recall, we are looking at the miles people drive to work. We do not consider distance to be negative so the lower inner fence and lower outer fence will not be part of our box plot. The box plot below was generated by https://plot.ly/. Be sure to label the fences appropriately.

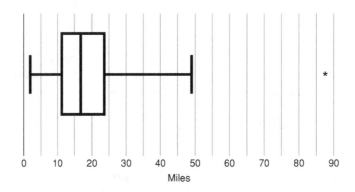

Critical thinking: Are there any outliers? If so, what kind? How do you know?

B. Identifying Outliers

One useful application of the box plot is in identification of outliers. As mentioned earlier, an outlier is a value in the dataset that is not quite in tune with other measurements. It is either much higher or much lower than all other values in the set. Identification of outliers is an important step in the process of data analysis because outliers could be very informative. Outliers could typically occur for a variety of reasons:

1. The measurement may have been incorrectly recorded. For example, when recording the value of the measurement, e.g., weight of individuals, if the decimal point is misplaced by one position, then clearly we would have an unusual measurement. As an example, if the weight of a person is 189.5 pounds and it recorded as 18.95, then this value would be an outlier and should be identified.
2. The apparatus with which the measurements are made might have become faulty. In manufacturing processes, the quantity of a variable or the amount of a substance may routinely be measured for use in the next stage of the process. If the apparatus that is used to make these measurements becomes faulty, final values will become outliers. It is clearly important to identify outliers to take corrective measures.

3. There may genuinely be an outlier in the measurements. From time to time unusual observations may occur in experiments. For example, if a doctor is reviewing the SBP of patients and encounters a value of say 195, which is an unusually high number for SBP, then it is important to notice this value and treat the respective patient.

We can therefore see how important it is to be able to identify outliers and go back and study them to find out the cause of occurrence of the outlier.

Generally, any observation falling beyond the inner fences can be considered to be problematic and may be a potential outlier. Thus if the value of an observation is less than or higher than it is an outlier and therefore must be scrutinized. However if the value is between the inner and outer fences, we call it a **mild** outlier. But, if the value of the observation is beyond the outer fences, we call it an **essential** or an **extreme** outlier. Thus an essential outlier is that which is either less than or higher than. Outliers are usually shown by asterisks on the boxplot.

■ Example 22: Commute to work—Outliers

In our commuting miles example, we can see that there are two potential outliers, namely 40 and 88 since both of these numbers are higher than the inner fence of 34. Since the data value 40 is below upper outer fence it is a **suspected outlier**. Since the data value 88 is above the outer fence it is an **outlier**. Generally, any observation falling beyond the inner fences can be considered to be problematic and may be a potential outlier. Thus if the value of an observation is less than(insert equation 1 here) or higher than (insert equation 2 here) it is an outlier and therefore must be scrutinized. However if the value is between the inner and outer fences, we call it a **mild** outlier. But, if the value of the observation is beyond the outer fences, we call it an **essential** or an **extreme** outlier. Thus an essential outlier is that which is either less than (insert equation 3 here) or higher than (insert equation 4 here). Outliers are usually shown by asterisks on the boxplot.

YOUR TURN! Please complete the required exercise below directly in this book. You are encouraged to discuss each exercise with a partner or group. You may be asked to complete the exercise in class or outside of class.

Exercise 15

In a given day, 16 adults came to doctor's office for care. The nurse measured the SBP for each patient and recorded the following:

123, 118, 137,128, 114, 104,185, 134, 141, 108, 112, 116, 122, 129, 135, 168

Calculate the following

Calculate the 5 number summary and the IQR.
Calculate the inner fences of the data set: $Q_3 + (1.5 * IQR)$ and $Q_1 - (1.5 * IQR)$.
Calculate the outer fences of the data set: $Q_3 + (3 * IQR)$ and $Q_1 - (3 * IQR)$.

Draw a boxplot and identify all outliers. Mark the outliers on your boxplot with asterisks and determine whether they are mild or essential (extreme) outliers.

Chapter 3

Probability

In Chapter 1, we defined statistical inference as the branch of statistics where we use the information in the sample to learn about the population and make statements about its unknown characteristics. In addition, statistical inference is also concerned about evaluating the reliability of these statements. So, how do we measure reliability? In order to measure reliability, we use probability to assess how well our statements can describe the characteristics of the population. In this chapter, we begin discussing probability and learn the basic concepts. We will continue our discussion with applications of probability in the next few chapters. Probability is a science that has grown significantly in the last quarter century. Many people specialize in different aspects of probability and there are several scholarly journals devoted to publishing theoretical and applied results in this important topic. Here, our discussion will focus on some elementary concepts of probability and we cover these concepts to an extent that will give us sufficient background for understanding statistical inference.

Section 1: Basic Definitions

In this section, we present the definition of some terms that are used to introduce probability concepts. We first define the notion of an experiment.

We define an **experiment** in its most general term as any process that leads to observation of one or more outcomes.

This process may be as simple as counting the number of words in a line of a manuscript or asking the opinion of your friend about a certain political action or could be as complex as observing the reaction of a substance to a specific chemical or making an observation through a mission in the moon or another planet. An experiment may be performed in a laboratory or any other place for that matter. There are, however, two types of experiments:

A: Experiments whose outcomes can be determined for sure and with certainty before the experiment is performed. For example, if you leave a rock from the top of a building, you know that by laws of gravity the rock will fall and hit the ground. We call this type of experiments **deterministic**.

B: Experiments whose outcomes cannot be predicted with certainty. Most of the experiments that we perform in practice are of this type. We do experiments because there are unknowns in the nature. To learn about these unknowns we use experimentation. We design and perform complex scientific experiments to learn about physical, biological, and chemical processes. We use cognitive experiments to explore psychological, sociological, and economical behaviors. This type of experiments is referred to as **nondeterministic** or simply **random** experiments. The terms probabilistic and stochastic have also been used for this type of experiments.

Because deterministic experiments are seldom of interest, we will not discuss them here and therefore every time we use the term 'experiment,' we mean the nondeterministic experiment. For these experiments, although we cannot exactly determine what the outcome is, often we can make a list of all possible outcomes. This list is called the **sample space** of the experiment and is usually denoted by S.

Any subset of the sample space is called an **event** and are usually denoted with capital letters of the alphabet such as A, B, C, and so on. Because we have defined event to be ANY subset of the sample space, an event may contain any number of outcomes from the sample space. If an event consists of only one outcome, then it is called a '**simple**' event. An even with no outcomes is a '**null**' event and the event with every outcome of the sample space is called the '**sure**' event.

Example 1: Rolling One Die

If we roll a regular die and count the number of pips (dots) on the side which the die lands, then,

$$S = \{1,2,3,4,5,6\}$$

represents the sample space. As you can see, there are 6 elements in this set. Now suppose that we define the following events: $A = \{1, 3, 5\}$, $B = \{4\}$ and $C = \{1, 2, 4, 5\}$.

If we look at event A, we could say, "A is defined as rolling an odd number." So, if we rolled a 4, we would say that event A did not occur. If we rolled a 1, would say that event A did occur. This idea holds for any event! If we roll a 4, we would say, "Event B did occur." If we rolled a 6, we would say that none of the events occurred.

Example 2: Flipping a Tail

Suppose we flip a coin until it lands on T (tail) for the first time and then we stop. The sample space of that experiment is

$$S = \{T, HT, HHT, HHHT, HHHHT, HHHHHT, ...\}$$

that is any sequence of H's (head's) ending with a T would be a member of this set. In practice, this model is applied when sampling is continued until the observation of a certain outcome.

The event that an odd number of flips of the coin will be necessary to terminate the experiment would be described as

$$D = \{T, HHT, HHHHT, HHHHHHT, ...\}$$

Example 3: Flu Vaccine

Now, suppose that a public health official selects two persons at random. Let V denote the outcome that the person is vaccinated against flu epidemic and N the outcome that the person is not, then there four possible outcomes in the sample space,

$$S_{\text{two people}} = \{VV, VN, NV, NN\}$$

Note the similarity of this model to tossing a coin twice. If we extend this experiment to three individuals rather than two, then there are eight outcomes,

$$S_{\text{three people}} = \left\{ \begin{array}{cccc} VVV & VVN & VNN & NNN \\ & VNV & NVN & \\ & VVN & NNV & \end{array} \right\}$$

Now, we can represent the outcomes in a sample space graphically using a tree diagram. In a tree diagram, we start at a point called a 'node' of the tree and use lines called branches to show the outcomes that follow, for example, the tree diagrams for the above example are as follows:

Tree Diagram

A tree diagram is often used to represent sample spaces graphically. A tree diagram consists of a set of nodes and branches. It starts from a point called the **root node** and branches are used to show the possible outcomes of the root. The process is continued until all steps of the experiment are represented. Each succession of the tree is called a **generation**. The following two graphs illustrate the tree diagrams for Example 3 above for the two cases when the health official selects two or three people to find out whether or not they are vaccinated against the flu epidemic. Note the first diagram consists of two generations of branches while the second diagram consists of three generations.

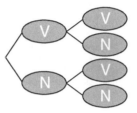

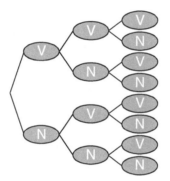

■ Example 4: Rolling Dice

An experiment consists of rolling a die twice. The sample space of this experiment consists of 36 outcomes:

$$
S = \left\{
\begin{array}{cccc}
(1,1) & (1,2) & (1,3) & \ldots (1,6) \\
(2,1) & (2,2) & (2,3) & \ldots (2,6) \\
\cdot & \cdot & \cdot & \cdot \\
\cdot & \cdot & \cdot & \cdot \\
\cdot & \cdot & \cdot & \cdot \\
(6,1) & (6,2) & (6,3) & \ldots (6,6)
\end{array}
\right\}
$$

QUESTION: List the events that correspond to the following:

Event A: Double—the same side both times

Event B: Total of 7

Event C: Total of 8

SOLUTION:

$A = \{(1,1)\ (2,2)\ (3,3)\ (4,4)\ (5,5)\ (6,6)\}$
$B = \{(1,6)\ (2,5)\ (3,4)\ (4,3)\ (5,2)\ (6,1)\}$
$C = \{(2,6)\ (3,5)\ (4,4)\ (5,3)\ (6,2)\}$

Critical Thinking:

○ *Suppose we look at the event, D, 6 on the first roll. List the outcomes.*

$D =$

○ *Suppose we look at the event E, a total of 11. List the outcomes.*

$E =$

○ *Suppose we look at the event, F, the sum is less than 6. List the outcomes.*

$G =$

YOUR TURN! Please complete the required exercises below directly in this book. You are encouraged to discuss each exercise with a partner or group. You may be asked to complete these exercises in class or outside of class.

Directions: There are five exercises below. Use the space provided at the bottom of the page and for each exercise

    a. Give the sample space.
    b. Give a tree diagram.

Exercise 1: A ball is drawn from a bag containing 10 balls numbered one through ten.

Exercise 2: A couple is planning to have three children. Let B denote the outcome of having a boy and G the outcome of a girl.

Exercise 3: In a manufacturing process, items coming off an assembly are examined one at a time until a defective item is found. Let N and D denote the outcome that the item is non-defective and defective, respectively.

Exercise 4: A die is rolled. If the outcome is an even number, a coin is flipped. Otherwise, if the outcome is an odd the die is rolled one more time.

Exercise 5: A bag contains three red, two white, four blue and one yellow marbles. Two marbles are drawn at random.

Exercise 1 Continued: In **Exercise 1**, write the outcomes in the following events:

A—A ball with an even number is drawn out of the bag:

$$A = \{ \qquad \}$$

B—A number that is a multiple of three is drawn:

$$B = \{ \qquad \}$$

C—A number that is a multiple of five is drawn:

$$C = \{ \qquad \}$$

Exercise 2 Continued: In **Exercise 2**, write the outcomes in the following events:

D—The family will end up with no boys:

$$D = \{ \qquad \}$$

E—The family will have exactly two boys:

$$E = \{ \qquad \}$$

F—The family will have at least one boy:

$$F = \{ \qquad \}$$

Exercise 3 Continued: In **Exercise 3**, write the outcomes in the event A that an odd number of items will be examined until a defective one is found:

$$A = \{ \qquad \}$$

Section 2: The Rules of Probability

Everyone knows that when we flip a fair coin in the air then there is a 50% chance that the coin will land on its heads side. Precisely, what is the meaning of this statement? Does this mean that if we flip the coin twice, the coin will land on each side once or if I flip a coin 20 times it will land on each side ten times? Clearly that is not the case. What it means is that if we flip the coin a large number of times, say 10,000 times, then we would expect the coin to land on each side about half the times. In fact we say that the probability that the coin lands on its heads side is 0.5.

Probability is the science that measures uncertainty in events. The measure of uncertainty in any event is, of course, important information and influences all decision making processes. As we discussed in the last section, we can associate a sample space to any experiment. Here, we see how we assign probabilities to all **simple events** in the sample space and use them to assess the probability of occurrence of any event. First we state the rules for probability. When we assign probabilities to simple events in the sample space, two rules must be satisfied:

✔ Rule 1: The probability of each simple event is a non-negative number.
✔ Rule 2: The sum of the probabilities assigned to all simple events in the sample space is equal to 1.

Therefore if our sample space consists of n outcomes say

$$S = \{a_1, a_2, \ldots an\}$$

Then in mathematical notations, Rule 1 states that $P(\{ai\}) \geq 0$ for $i = 1 \; 2 \ldots n$ and Rule 2 states that $\sum_{i=1}^{n} P(\{a_i\}) = 1$.

The assignment of probabilities are done in such a way that the probability of each simple event represents its likelihood of occurrence and it is, as we mentioned before, the long term proportion of times that the outcome is expected to occur in a large number of replications of the experiment.

For example, if the experiment consists of tossing a fair coin and H and T respectively denote the events of the coin landing on heads and tails, we assign $P(H) = \frac{1}{2}$ and $P(T) = \frac{1}{2}$. But, if we know that the coin is biased and lands on H twice as many times that it lands on T, then we assign $P(H) = \frac{2}{3}$ and $P(T) = \frac{1}{3}$.

■ Example 5: Rolling Dice

An experiment consists of rolling a die and observing the number of pips on the up side of the die. In this experiment, the sample space has six outcomes,

$$S = \{1, 2, 3, 4, 5, 6\}$$

QUESTION: Which of the following probability assignments is legitimate?

a. $P(\{1\}) = P(\{2\}) = P(\{3\}) = P(\{4\}) = P(\{5\}) = P(\{6\}) = 1/6$

b. $P(\{1\}) = 1/3, P(\{2\}) = 0, P(\{3\}) = 1/6, P(\{4\}) = 0, P(\{5\}) = 1/3, P(\{6\}) = 1/6$

c. $P(\{1\}) = 1/6, \; P(\{2\}) = 1/3, \; P(\{3\}) = 0, \; P(\{4\}) = 0, \; P(\{5\}) = 1/12, \; P(\{6\}) = 1/3$

d. $P(\{1\}) = 1/6, \; P(\{2\}) = 1/3, P(\{3\}) = 1/6, \; P(\{4\}) = 1/12, \; P(\{5\}) = 1/3, \; P(\{6\}) = 0$

REMEMBER: The two rules for probability assignment must be satisfied in order to make that assignment legitimate.

SOLUTION:

○ In the above four cases, we can easily see that part a refers to a balanced die and not only all sides have the same probability, the probabilities add up to 1. Therefore it is legitimate.

○ Part b refers to a die that has numbers 1and 5 each on two sides and numbers 3 and 6 each on one side. The probabilities add up to one and the assignment is legitimate.

○ In parts c and d, however, although each probability is a non-negative number, the sum of probabilities is not 1. Therefore the probability assignment is not legitimate.

■ Example 6: Rolling Dice

A die is loaded in such a way that each side with an even number of pips is twice as likely as the sides with odd number of pips. If this die is rolled once, find the probability of each side.
Here, again, the sample space is

$$S = \{1, 2, 3, 4, 5, 6\}$$

And we note that in this case, the die is not balanced and therefore the probabilities for all sides are not equal. Rather, the sides with an even number of pips i.e., 2, 4, and 6 are twice as likely as the sides with an odd number of pips, i.e., 1, 3, and 5.

QUESTION: What is the probability for each number of the sample points of ?

SOLUTION: Since we ONLY know how the two probabilities are related (and the not the actual probabilities themselves), let the probability that a single roll of the die lands on 1 be denoted by p, therefore we have. Since the probability of sides with an even number of pips is twice the probability of the sides with an odd number of pips, we have:

$$P(\{1\}) = p \qquad P(\{2\}) = 2p$$
$$P(\{3\}) = p \qquad P(\{4\}) = 2p$$
$$P(\{5\}) = p \qquad P(\{6\}) = 2p$$

Now, since according to the laws of probability the sum of the probabilities of all simple events in the sample space must be equal to 1, we have

$$p + 2p + p + 2p + p + 2p = 1$$
$$9p = 1$$
$$p = \frac{1}{9}$$

The algebra above gives us an answer of $p = 1/9$. What does this answer mean? Recall we defined p to be the probability of rolling an odd number, specifically, rolling a 1. Since rolling an **even number is twice as likely** as rolling an odd number, the probability

$$P(\{2\}) = P(\{4\}) = P(\{6\}) = 2 * \frac{1}{9} = \frac{2}{9}$$

Exercise 6: Consider a ten-sided die and assign probabilities to each side if:

a. All sides have the same chance

$P(1) =$ $P(2) =$ $P(3) =$ $P(4) =$ $P(5) =$

$P(6) =$ $P(7) =$ $P(8) =$ $P(9) =$ $P(10) =$

b. Each side with an even number is three times as likely as each side with an odd number.

$P(1) =$ $P(2) =$ $P(3) =$ $P(4) =$ $P(5) =$

$P(6) =$ $P(7) =$ $P(8) =$ $P(9) =$ $P(10) =$

Exercise 7 (Continued): In Exercise 2, you were asked to write the sample space for a couple who plans to have three children. If the chance of giving birth to a boy is the same as the chance of giving birth to a girl, what probability would you assign to each simple event in this sample space?

Critical Thinking: *If in Exercise 2, the couple plans to have four children, how many outcomes do you think the sample space would have? If boys and girls have equal chance of birth, what probability would you assign to each simple event in that sample space?*

Section 3: Probability of an Event

Now that we have set the rules for probability assignment to simple events in the sample space, we can describe how we can find the probability of any event.

If A is an event, then its **probability**, denoted by $P(A)$, is defined as the sum of the probabilities of the simple events contained in A.

■ Example 7: Rolling Dice

In a previous example, we computed the probability for each side of a die that is not balanced and is loaded in such a way that each side with an even number of pips is twice as likely as each side with an odd number of pips. Suppose now this die is rolled once and consider the events A and B defined as

$$A = \{1,2,3\}, \quad B = \{2,5\}$$

QUESTION: What are the probabilities of the events above?

SOLUTION: In order to find the probability of each of these two events, we need to add the probabilities of the simple events contained in each event. Consequently, we have

$$P(A) = P(\{1\}) + P(\{2\}) + P(\{3\}) = \frac{1}{9} + \frac{2}{9} + \frac{1}{9} = \frac{4}{9}$$

$$P(B) = P(\{2\}) + P(\{5\}) = \frac{2}{9} + \frac{1}{9} = \frac{1}{3}$$

Critical thinking: What is the value of $P(A) + P(B)$? Discuss your answers with your classmates.

■ Example 8: Rolling Dice with Different Colors

A pair of balanced dice, a red die and a blue die, is rolled.

QUESTION: What are the probabilities of the events listed below?

    A: Double (outcomes are the same on both dice)
    B: Sum of the number of pips on both dice is 7
    C: Sum of the number of pips on both dice is 8
    D: The red die lands on the side with 6 pips

SOLUTION: We considered a similar example earlier in Example 4. However, in that example we said that a die is rolled twice, but here we are rolling a pair of dice, say a blue die and a red die.

We saw that the sample space contained 36 possible outcomes with each outcome expressed as an ordered pair, the first number representing the outcome of the first roll and the second outcome representing the outcome of the second roll. Note that if in each ordered pair we designate the first number as the outcome of the red die and the second number as the outcome of the blue die, we have the same sample space as in Example 4.

Since the two dice are fair, it makes sense to think that all 36 outcomes have the same likelihood of occurrence and therefore we assign a probability of 1/36 to each simple event in the sample space, that is

$$P(\{i, j\}) = \frac{1}{36} \quad for\ i = 1, 2, \ldots, 6\ and\ j = 1, 2, \ldots, 6.$$

In other words the outcomes in this experiment are all **equally likely**.

Now, to find the probability of the events A, B, C, D, we simply need to add the probabilities of the respective simple events contained in that event.

$$P(A) = P(\{1,1\}) + P(\{2,2\}) + P(\{3,3\}) + P(\{4,4\}) + P(\{5,5\}) + P(\{6,6\})$$

$$= \frac{1}{36} + \frac{1}{36} + \frac{1}{36} + \frac{1}{36} + \frac{1}{36} + \frac{1}{36}$$

$$= \frac{6}{36}$$

$$= \frac{1}{6}$$

$$P(B) = P(\{1,6\}) + P(\{2,5\}) + P(\{3,4\}) + P(\{4,3\}) + P(\{5,2\}) + P(\{6,1\})$$

$$= \frac{1}{36} + \frac{1}{36} + \frac{1}{36} + \frac{1}{36} + \frac{1}{36} + \frac{1}{36}$$

$$= \frac{6}{36}$$

$$= \frac{1}{6}$$

Similarly we have,

$$P(C) = P(\{2,8\}) + P(\{3,5\}) + P(\{4,4\}) + P(\{5,3\}) + P(\{6,2\}) = \frac{5}{36}$$

$$P(D) = P(\{6,1\}) + P(\{6,2\}) + P(\{6,3\}) + P(\{6,4\}) + P(\{6,5\}) + P(\{6,6\}) = \frac{1}{6}$$

We see from the above example that because the outcomes in the sample space are equally likely, there seemed to be an easier way to calculate the probability. It appears that each time all we have to do is to count the number of outcomes in the event and divide that by the number of outcomes in the sample space. Therefore, we can say that if the outcomes in the sample space are equally likely, then the probability of any event A is given by

$$P(A) = \frac{(\textit{Number of outcomes in } A)}{\textit{Number of outcomes in the sample space } S}$$

We must remember that **this rule only applies to sample spaces in which the outcomes are equally likely**.

In some cases, even though the sample space *does not* contain equally likely outcomes, we may be able to rewrite it in such a way that it does contain equally likely outcomes.

■ Example 9: A Bag of Marbles

A bag contains three white and two red marbles. Two marbles are drawn at random and without replacement.

QUESTION: What are the probabilities of the events listed below?

Event A: No red marbles

Event B: One red marble

Event C: Two red marbles

SOLUTION: The sample space of this experiment can simply be expressed as

$$S = \{WW, WR, RW, RR\}$$

where W denotes drawing a white marble and R denotes drawing a red marble. The tree diagram is also as follows:

WARNING! In this sample space clearly the outcomes are not equally likely due to the fact the number of white and red marbles in the bag is not the same.

   Intuitively we know that drawing two white marbles has a higher chance than drawing two red marbles since there are three white marbles in the bag compared to only two red marbles. We can write the sample space in such a way that it would account for this difference in the number of white and red marbles. Since there are three white marbles, we call them W_1, W_2, and W_3. Similarly we number the two red marbles as R_1 and R_2. We can then express the sample space as:

$$S = \{W_1\ W_2 \quad W_1\ W_3 \quad W_1\ R_1 \quad W_1R_2$$
$$W_2\ W_3 \quad W_1\ R_1 \quad W_2R_2$$
$$W_3R_1 \quad W_3R_2$$
$$R_1R_2\}$$

Here $W_1\ W_2$ denotes the outcome that the white marbles numbered 1 and 2 are drawn out of the bag in any order and so on.

   Now the sample space has ten equally likely outcomes as it accounts for the fact that there are more white marbles than red marbles. Therefore we can assign a probability of 0.1 to each simple event.

   Solving the problem now becomes rather trivial. In the expanded sample space, there are three outcomes that correspond to drawing two white marbles and no red marbles out of the bag. Thus $P(no\ red\ marbles) = 0.3$. Similarly, there are six outcomes corresponding to one white and one red marbles being drawn and therefore $P(one\ red\ marbles) = 0.6$. Finally, there is only one outcome relating to the event of drawing two red marbles out of the bag and hence $P(two\ red\ marbles) = 0.1$

   We can see that by rewriting the sample space in this expanded format so that all outcomes are equally likely, we were able to solve this problem. It should be mentioned, however, that this method can reasonably work if the total number marbles is relatively small and if the number of marbles in the bag is large, then writing the expanded sample space is not feasible. We will return to this problem later and will discuss another method of solving this problem in which the size of the population really does not matter.

> **YOUR TURN!** Please complete the required exercises below directly in this book. You are encouraged to discuss each exercise with a partner or group. You may be asked to complete these exercises in class or outside of class.

Exercise 7: Consider a ten-sided die. If this die is rolled, find the probability of the events

Event A: The die lands on a side with an even number of pips

Event B: The die land on 1 or 10

Event C: The die lands on 2, 5, 7, or 8

There will be two cases, in each of those cases, find the probability of the event. SHOW ALL WORK!

Case I: All sides have the same chance.

$P(A) =$

$P(B) =$

$P(C) =$

Case II: Even numbers are three times as likely as odd numbers.

$P(A) =$

$P(B) =$

$P(C) =$

Exercise 8: In a small seminar class, there are three seniors, two juniors and one sophomore. Two students are to be selected for a special demonstration.

a. Number the seniors as S_1, S_2, and S_3, the juniors as J_1 and J_2, and the sophomore as P. Write a sample space with equally likely outcomes that accounts for the fact that the number of student from each year of study is different in the class:

S = {   }

b. If these two students are selected at random, compute the probability that the number of seniors in the sample is

$P(no\ seniors) =$

$P(one\ senior) =$

$P(two\ seniors) =$

Section 4: Union and Intersection of Two Events

Thus far we have considered the properties of single events from different sample spaces. Here we discuss combination of events. If we have two events from a sample space, we define new events that somehow combine these two events.

Suppose A and B are two events:

The **union** of A and B denoted by $A \cup B$ is the event that either A occurs or B occurs or both. It contains all outcomes that are either in A or in B or both.

The **intersection** of A and B denoted by $A \cap B$ is the event that A and B occur simultaneously together. It contains all outcomes that are in common between the two events.

■ Example 10: Bag of Marbles—Continued

A bag contains ten identical marbles with numbers 1 through 10 written on them. One marble is drawn at random. Let the events A, B, and C be defined as follows:

Event A: an even number is drawn

Event B: a multiple of three is drawn

Event C: a multiple of five is drawn

In this example, we have

$$A = \{2,4,6,8,10\}, \qquad B = \{3,6,9\} \ and \ C = \{5,10\}$$

Therefore,

$$A \cup B \ = \ \{2,3,4,6,8,9,10\}$$

This means that if any of the numbers in $A \cup B$ is drawn, then it is either an even number or a multiple of 3 or both. Also, since 6 is the only outcome in common between A and B,

$$A \cap B = \{6\}$$

which again means that if the marble with number 6 is drawn out of the bag, then it is both an even number and a multiple of 3. Similarly,

$$A \cup C = \{2,4,5,6,8,10\}, \ A \cap C = \{10\}$$

$$B \cup C = \{3,5,6,9,10\}, \quad B \cap C = \phi(the \ empty \ set)$$

Note that in the above, the intersection of the two events B and C is the empty set since the two events have no outcomes in common and they cannot occur together. We call these events disjoint or mutually exclusive. Two events are called **disjoint** or **mutually exclusive** if their intersection is the empty set.

Section 5: The Complement of an Event

If A is an event, then its **complement**, denoted as $\bar{A}$ or A^c is the event that A does not happen. It contains all elements of the sample space that are **not included in A**.

■ Example 11: Bag of Marbles—Continued

A bag contains ten identical marbles with numbers 1 through 10 written on them. One marble is drawn at random. Let the events A, B, and C be defined as follows:

Event A: an even number is drawn

Event B: a multiple of three is drawn

Event C: a multiple of five is drawn

In this example, we have

$$A = \{2,4,6,8,10\}, B = \{3,6,9\} \text{ and } C = \{5,10\}$$

QUESTION: What elements are in the following sets:

$$A^c, (A \cup B)^c, (B \cap C)^c \text{ and } (A^c \cup B)^c ?$$

SOLUTION:

■ Here, since A contains all the even numbers that are in the sample space, A^c would consist of the odd numbers, that is

$$A^c = \{1,3,5,7,9\}$$

■ Since $A \cup B = \{2,3,4,6,8,9,10\}$ the complement would be all numbers NOT in $A \cup B$. Hence $(A \cup B)^c = \{1,5,7\}$.

■ Since $B \cap C = \phi$ the complement would be all numbers NOT in $B \cap C$. Hence $(B \cap C)^c = \{1,2,3,4,5,6,7,8,9,10\}$

■ In order to find $(A^c \cup B)^c$ we will need two steps. First we find $A^c \cup B$.

$$A^c \cup B = \{1,3,5,7,9\} \cup \{3,6,9\} = \{1,3,5,6,7,9\}$$

then we look to see what elements are not listed and now we have our answer:

$$(A^c \cup B)^c = \{2,4,8,10\}$$

Critical Thinking: For the above example, what is $A \cap A^c$? Discuss and explain.

YOUR TURN! Please complete the required exercise below directly in this book. You are encouraged to discuss each exercise with a partner or group. You may be asked to complete the exercise in class or outside of class.

Exercise 9: One card is drawn at random from a standard 52-card deck of playing cards. Let C be the event that the suit of the card is heart, D be the event that the card is a face card (Jack, Queen, and King), and E be the event that the card is either a King or an Ace. What outcomes are there in the following events?

a. $D \cap C$

b. $D \cup C$

c. D^c

d. $D \cap C^c$

e. $E \cap D \cap C$

f. $(E \cup D)^c \cap C$

Section 6: Pictorial Representation

The concept of union, intersection, and complement can also be graphically represented using a Venn diagram. In a Venn diagram, we represent the sample space with a rectangle and show the events with circles inside the rectangle. Therefore an event A will be represented as follows:

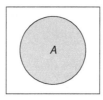

Figures below show the Venn diagrams for union, intersection of the two events A and B together with the complement of A.

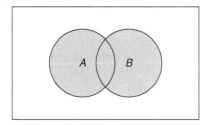

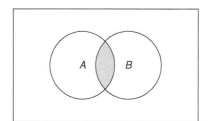

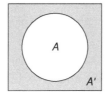

Addition Law of Probability

How do we find the probability of the union of two events? Is there a relationship between the probability of the intersection of two events and the individual probabilities of the two events? Well, let us try to answer this question by looking at the Venn diagram. In the Venn diagram showing the union of two events, we can see that the probability of the union of two *events A and B can expressed as the sum of the probabilities of the two events A and B. However in adding the probability of A to probability of B, we can see that in fact the section that is double counted is the intersection of A and B. Hence, to compensate for it, we subtract this probability once. Thus we have*

$$P(A \cup B) = P(A) + P(B) - P(A \cap B),$$

which is called the **additive rule** of probabilities.

■ Example 12: Bag of Marbles—Continued

Previously we defined events A and B as

$$A = \{2, 4, 6, 8, 10\} \text{ and } B = \{3, 6, 9\}$$

and found that $P(A) = 0.5$ and $P(B) = 0.3$. We also found that $P(A \cup B) = 0.7$ and $P(A \cap B) = 0.1$.

We will now calculate the value of $P(A \cup B)$ in a different way, using the additive rule!

$$P(A \cup B) = P(A) + P(B) - P(A \cap B) = .5 + .3 - .1 = .7$$

■ Example 13

On her way to work, Martha has to go through two intersections. From past experience Martha knows that the chance having to stop at each intersection because of red lights is 50%. She also has found that in 20% of cases she had to stop at both intersections. What is the probability that Martha would have to stop at least once on her way to work? In what percent of cases can she drive straight through without stopping?

Let A_1 and A_2 be the events that Martha would have to stop at the first and the second intersections respectively. Then we have

$$P(A_1) = P(A_2) = 0.5$$

Also

$$P(A_1 \cap A_2) = 0.2$$

Thus the probability that at least one of the two events A_1 or A_2 occurs, i.e., their union, by the additive rule, is given by

$$P(A_1 \cap A_2) = 0.5 + 0.5 - 0.2 = 0.8$$

That is to say that in 80% of cases Martha has to stop at least once. Therefore, the event that she does not have to stop at all is the complement of the union of the two events, that is

$$P(A_1 \cup A_2) = 1 - 0.8 = 0.2$$

Special Case

Note that if two events B and D are mutually exclusive, since the intersection of the two events is the empty set, as illustrated in the Venn diagram below, the additive rule becomes

$$P(B \cup D) = P(B) + P(D)$$

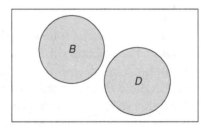

which is the **additive rule for disjoint events.** Therefore, for two disjoint events, the probability of their union is simply the sum of their individual probabilities. This is also evident from the Venn diagram.

YOUR TURN! Please complete the required exercises below directly in this book. You are encouraged to discuss each exercise with a partner or group. You may be asked to complete these exercises in class or outside of class.

Exercise 10: In a city, it is known that 80% of homes have at least of one TV set in the house. Also, 85% of homes have internet services. If 75% have both, at least one TV set and internet services, what is the probability that a randomly selected home in this city will have at least one TV set or internet services?

Here if A is the event that a randomly selected home would have at least one TV set, then

$P(A) =$

Similarly, if B is the event that the selected home would have internet services, then

$P(B) =$

Now the probability that the selected home would have both, at least one TV set and internet services is 0.75. Thus

$P(A \cap B) =$

Hence by the additive rule the probability that the selected home would have at least one TV set or internet services is

$P(A \cup B) =$

Exercise 11: If 40% of students in a class are in Math 101 and 58% are in Spanish 100 and that 23% are taking both Math 101 and Spanish 100, what percent of students are either in Math 101 or Spanish 100?

$P(Math \cup Spanish) =$

Section 7: Conditional Probability

In the last section, we learned how to find probability of events. Often, we may have some partial information that we may wish to incorporate in the evaluation of probability. To understand the concept, let us use a simple example. Suppose an experiment consists of rolling a balanced die once and observing the up side of the die. Then we know that the sample space of this experiment is given by

$$S = \{1, 2, 3, 4, 5, 6\}$$

Let the event A be defined as the event that the up side of the die has an even number of pips, that is

$$A = \{2, 4, 6\}$$

Then, clearly as we learned in the last section, $P(A) = 3/6 = \frac{1}{2}$. Suppose now that we are told that the number of pips on the up side of the die is a number higher than three. Although we do not have complete information about the outcome, we have some partial information that clearly influences the probability of A. Given this information, we note that the probability of A is in fact no longer $\frac{1}{2}$. Since there are three numbers higher than three that are possible and two of them are higher even numbers, the probability of occurrence of A is 2/3. Therefore conditional on the number of pips on the up side of the die being higher than three, the probability of it being even is 2/3. If we denote by B the event that the number of pips on the up side of the die is higher than three, we say the conditional probability of A given B is 2/3 and we write

$$P(A|B) = 2/3.$$

Conditional probability is quite important with many applications. In practice, very often we get new information about an event that can change the probability of the event. For example, in the court of law as new evidence such as the blood type at the crime scene, the DNA results, and so on comes in, the jury changes the chances of conviction. We formally define conditional probability of A given B as follows:

$$P(A|B) = \frac{P(A \cap B)}{P(B)}$$

■ Example 14

In the die example that we just described, we defined the events A and B as

$$A = \{2, 4, 6\} \qquad \text{and} \qquad B = \{4, 5, 6\}$$

Therefore

$$P(A|B) = \frac{P(A \cap B)}{P(B)} = \frac{2/6}{3/6} = 2/3$$

which is what we found using our intuition.

■ Example 15: (Cancer and Smoking)

In an epidemiologic study to investigate the relationship between smoking and dying of cancer, a random sample of 500 individuals from a certain population was selected and individuals were observed over a long period of time (longitudinal study). After death, individuals were classified according to two attributes:

a. Whether the person was a smoker or nonsmoker

b. Whether the person died of cancer or other causes

The following two-way table was resulted:

	Smoker	Nonsmoker	Total
Died of cancer	98	22	120
Died of other causes	42	338	380
Total	140	360	500

If an individual is selected at random from this population, what is the probability that he/she:

a. Is a smoker?

b. Dies of cancer?

c. Dies of cancer and is a smoker?

d. Dies of cancer or is a smoker?

e. Dies of cancer if the person is a smoker?

f. Dies of cancer if the person is a nonsmoker

g. Dies of other causes if the person is a smoker?

Here, we can use the data in the table to estimate these probabilities. We note that out the 500 individuals a total of 140 were smokers. Therefore for part a.

$$P(\text{Smoker}) = \frac{140}{500} = 0.28$$

Similarly, for part b., we have

$$P(\text{Cancer}) = \frac{120}{500} = 0.24$$

Now, for part c., we notice that there were 98 people who were both a smoker and died of cancer. Thus

$$P(\text{Cancer and Smoker}) = \frac{98}{500} = 0.196$$

To answer part d., we apply the additive rule

$$P(\text{Cancer or Smoker}) = P(\text{Cancer}) + P(\text{Smoker}) - P(\text{Cancer and Smoker})$$

$$= 0.24 + 0.28 - 0.196 = 0.324$$

In part e., we are looking for the conditional probability of dying of cancer given the individual is a smoker. Knowing that the individual is a smoker, what is the chance that he/she would die of cancer? Using the conditional probability formula, we have

$$P(\text{Cancer}\,|\,\text{Smoker}) = \frac{P(\text{Cancer and Smoker})}{P(\text{Smoker})} = \frac{0.196}{0.28} = 0.70$$

Interestingly, we see how the knowledge of smoking status of the person has drastically changed the chance of dying of cancer. With no knowledge about smoking status, the chance of dying of cancer is 28%, i.e., it is estimated that about 28% of the population at large dies of cancer. However as soon as we find out that that person is a smoker that chance jumps to 70%. This is how conditional probability can be applied to update probability. Note also that by looking at the table of data, we could estimate the probability of dying of cancer given that the person is smoker in a more direct way. Since we know the person is a smoker, and among 120 smokers, 98 died of cancer, we can argue that

$$P(\text{Cancer}\,|\,\text{Smoker}) = \frac{98}{140} = 0.70,$$

which is the same result as before.

Similarly, for part f. of the problem we can write

$$P(\text{Cancer}\,|\,\text{Nonsmoker}) = \frac{22}{360} = 0.061$$

This again demonstrates how conditional probability can drastically change the chance of occurrence of an event. Here, we see that once we find out that the individual is not a smoker, the chance of dying of cancer drops from 28% to only 6.1%. Finally, for part g of this problem, we can find the conditional probability of dying of other causes given that the person is a smoker as

$$P(\text{Other causes}\,|\,\text{Smoker}) = \frac{42}{140} = 0.30$$

Note that the sum of $P(\text{Cancer}\,|\,\text{Smoker})$ and $P(\text{Other causes}\,|\,\text{Smoker})$ is equal to 1.0. This makes sense since if an individual is a smoker, then he/she either dies of cancer or other causes.

The above example clearly illustrates the importance of conditional probability. We can see how by using conditional probability we can incorporate any information in the evaluation of the chance of occurrence of an event. Applications of conditional probability occur in practice all the time. For example, a physician uses all the evidence such as the body temperature, the blood pressure, the lab results, and so on in order to make a diagnosis about the patient. The field of artificial intelligence in Computer Science uses conditional probability in order to develop software that helps physicians in their decision making regarding diagnosis of complex diseases. As another application, we can name the insurance industry. When insurance companies try to evaluate the premium for a policy, all the risk factors are used in order to determine the probability of an adverse event. For example, to determine the premium for an auto insurance policy, they use information on the age, driving history, previous accidents, county of residence, and so on of the driver in order to find the chance of having an accident. Based on that probability, they give the premium.

Exercise 12: A card is drawn at random from a standard deck. Find the probability that the selected card is

a. Heart

$$P(\text{Heart}) = \frac{13}{}$$

b. A face card

$$P(\text{Face}) = \frac{}{52}$$

c. Heart and face card

$$P(\text{Heart} \cap \text{Face}) = -$$

d. Heart or face card

$$P(\text{Heart} \cup \text{Face}) =$$

e. Heart given it is a face card

$$P(\text{Heart} \mid \text{Face}) = -$$

f. Heart and not a face card

$$P(\text{Heart} \cap \text{Not Face}) = -$$

g. Any suit but Heart given it is a face card

$$P(\text{Not Heart} \mid \text{Face}) = -$$

Multiplication Rule of Probabilities

In the last section, we learned the condition probability formula. If in that formula, we multiply both sides of the equation by $P(B)$, we get

$$P(A \cap B) = P(A \mid B) * P(B),$$

which is called the multiplication rule for probabilities. This formula can often be used to find probability of joint occurrence of events.

■ Example 16: City Generators

The electrical system of a city operates on two generators; the main generator and the backup. From past experience, we know that the probability that the main generator breaks down is only 0.0001. If the main generator breaks down, the backup automatically starts working. However, the backup is a weaker engine and has a 2 in 1000 chance of failing. We wish to find the probability that the entire electrical system shuts down.

Let M be the event that the main generator fails, then $P(M) = 0.0001$. Let B be the event that the backup fails to operate, then we know that if the main generator fails, the backup has a probability of 0.002 of failure, that is $P(B \mid M) = 0.002$. Now, using the multiplication rule, we have

$$P(B \cap M) = P(B \mid M) * P(M) = 0.002 * 0.0001 = .0000002 = 2 * 10^{-7}$$

and hence there is only a 2 in 10 million chance for the entire system to shut down.

■ Example 17: Marbles in a Bag

A bag contains three white and two red marbles. Two marbles are drawn at random and without replacement. We wish to find the probability of drawing

  a. Two white marbles

  b. Two red marbles

  c. One of each color

We saw this problem earlier in this chapter. We realized that in order to be able to calculate the above probabilities, we had to write a sample space with equally likely outcomes. That meant numbering the marbles and writing an extended sample space. As we commented at that point, the method of numbering the marbles may work if we have a small number of marbles of each color in the bag. However if the number of marbles in bag becomes rather large, then writing the extended sample space will be a daunting task and very time consuming. Here, we see how by using the multiplication rule, we can solve the same problem in a much easier fashion. We will see that in this method, the number of marbles in the bag does not really matter and the method can easily be applied irrespective of the number of marbles of each color in the bag.

Let W_1 and W_2 be respectively the events of drawing white marbles on the first and second draws. Then the probability of drawing two white marbles is, by the multiplication rule, given by

$$P(W_2 \cap W_1) = P(W_2 \mid W_1) * P(W_1)$$

Now, since there are three white marbles in the bag out of a total of five, $P(W_1) = 3/5 = 0.6$. If the first marble drawn is white, there only remains four marbles in the bag only two of which are white. This means that $P(W_2 | W_1) = 2/4 = 0.5$. Hence

$$P(W_2 \cup W_1) = 0.5 * 0.6 = 0.3$$

Similarly, for part b. of the problem, if we let R_1 and R_2 be respectively the events of drawing red marbles on the first and second draws, then we see that since there are two red marbles out of five, $P(R_1) = 2/5 = 0.4$. If the first marble drawn is red, the chance that the second one is also red is only one out of a total of four. That is $P(R_2 | R_1) = \frac{1}{4} = 0.25$. Thus again by applying the multiplication rule we have

$$P(R_2 \cap R_1) = P(R_2 | R_1) * P(R_1) = 0.25 * 0.4 = 0.1$$

Finally, for part c., we realize that there are two scenarios that could lead to drawing of one marble of each color. One scenario is to draw a white marble first and a red marble on the second drawing. The other scenario is to first draw a red marble followed by a white marble. This means that for part c. we need to apply the multiplication rule twice. Consequently, we have

$$P(\text{One of each color}) = P(W_2 \cap R_1) + P(R_2 \cap W_1)$$

$$= P(W_2 | R_1) * P(R_1) + P(R_2 | W_1) * P(W_1)$$

$$= \frac{3}{4} * \frac{2}{5} + \frac{2}{4} * \frac{3}{5} = \frac{12}{20} = 0.6$$

We note that these results are exactly the same as what we had when we solved this problem using the extended sample space. However, this method is much more amenable. We also note that we can obtain these results simply by using a tree diagram. The tree diagram below depicts the situation for this problem. The number on each branch gives the probability for the event represented by that branch.

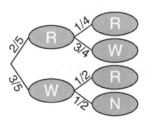

We can readily see that the branch of the tree that represents the probability of drawing two white marbles can occur with the probability of 0.5*0.6 = 0.3 and the branch representing the occurrence of drawing two red marbles has a probability of 0.4* 0.25 = 0.1. Also, we can see from the tree that the event of drawing one marble of each color can be shown by two branches of the tree leading to the same probability of 0.6 as before.

Exercise 13: The records of a basketball player show that in shooting two free throws, she can make 74% of her first shots. If she makes the first shot, she has a 82% chance of also making the second shot while if she does not make the first shot, she has a 65% chance of also not making the second shots. What is the probability that:

a. She makes both shots?

$$P(\text{making both shots}) = \text{_____}$$

b. She fails to make either of the shots?

$$P(\text{failing on both shots}) = \text{_____}$$

■ Section 8: Independent Events

In order to better describe the concept of independent events, we begin with an example.

■ Example 18: Coin Flip

A fair coin is flipped twice. We wish to calculate the probability that this coin would land on Heads on the second flip in two cases: Once when it is known that the coin has landed on Heads on the first flip and once with no knowledge of the outcome of the first flip. Let A_1 and A_2 be defined as follows:

A_1—The event the coin will land on Heads on the first flip

A_2—The event the coin will land on Heads on the first flip

Now, the sample space of this experiment can be expressed as:

$$S = \{HH, HT, TH, TT\}$$

Where, as usual, H and T respectively denote the outcomes of Heads and Tails in the flip of the coin. Thus the events A_1 and A_2 can be expressed as:

$$A_1 = \{HH, HT\} \qquad \text{and} \qquad A_2 = \{HH, TH\}$$

Note that the outcome in the sample space are equally likely and therefore $P(A_1) = \frac{1}{2}$ and $P(A_2) = \frac{1}{2}$. Moreover, since A_1 and A_2 have only one outcome in common, $P(A_1 \cap A_2) = 1/4$. Suppose first that we know the coin landed on Heads the first time it was thrown in the air, then the probability that it will also land on Heads the second time that is $P(A_2 | A_1)$ can be calculated using the definition of conditional probability as follows:

$$P(A_2 \,|\, A_1) = \frac{P(A_2 \cap A_1)}{P(A_1)} = \frac{1/4}{1/2} = \frac{1}{2}$$

Next, as mentioned above, the probability that the coin lands on Heads the second time it is flipped with no knowledge of the outcome of the first flip is:

$$P(A_2) = \frac{1}{2}$$

Thus we see that in this case, interestingly, the conditional probability of A_2 given A_1 is identical to the unconditional probability of A_2. This, of course, did not happen in the previous examples that we saw of conditional probability. In the die example, when we knew that the die landed on a number higher than three, the probability of it landing on as even number changed. Similarly in the cancer and smoking example, we saw that when we knew that the individual was a smoker, the probability of dying of cancer changed. But, in the present example, the knowledge of the outcome of the first flip does not change the probability of the coin landing on Heads the second time. The probability of the coin landing on Heads on the second flip is ½ whether or not we know the outcome of the first flip. In this case we say that the two events A_1 and A_2 are independent. Thus two events are independent when the conditional probability of one given the other is the same as the unconditional probability of that event.

Independent Events: Call two events A and B **independent** when

$$P(A|B) = P(A)$$

Note that if we replace the left side of the above equation with the definition of conditional probability, we have

$$\frac{P(A \cap B)}{P(B)} = P(A)$$

and upon multiplying both sides of this equation by $P(B)$, we obtain

$$P(A \cap B) = P(A) * P(B)$$

which is an equivalent definition of independent events, i.e., two events A and B are **independent** if the probability of their joint occurrence is equal to the product of their individual probabilities. This is also called the **multiplicative rule for independent events**.

■ Example 19: Rolling a die

A die is rolled twice. Let, as before, events A, B, C, and D be defines as:

   A—A double

   B—A total of seven

   C—A total of eight

   D—Six on the first flip

Are A and B independent? How about A and D? B and D?

   In this example, note that A and B have no outcomes in common and therefore the conditional probability of A given B is

$$P(A|B) = \frac{P(A \cap B)}{P(B)} = \frac{0}{1/6} = 0$$

whereas the unconditional probability of A,

$$P(A) = \frac{6}{36} = 1/6$$

and since the conditional probability is not the same as the unconditional probability, the two events A and B are not independent. Equivalently, we could say that $P(A \cap B) = 0$ and

$$P(A) * P(B) = \frac{1}{6} * \frac{1}{6} = \frac{1}{36}$$

and since the product of individual probabilities is not the same as the probability of the joint occurrence of the two events, A and B are not independent.

Now, consider the two events A and D. These two events have one outcome in common and that is (6,6). Therefore

$$P(A|D) = \frac{P(A \cap D)}{P(D)} = \frac{\frac{1}{36}}{\frac{1}{6}} = 1/6$$

And the unconditional probability of A, $P(A) = 1/6$. Therefore in this case the conditional probability of A is equal to the unconditional probability of A and hence the two events A and D are independent. Equivalently, we could argue that since $P(D) = 1/6$, we have

$$P(A \cap D) = \frac{1}{36} \quad \text{and} \quad P(A)P(D) = \frac{1}{6} * \frac{1}{6} = \frac{1}{36}$$

And because the probability of joint occurrence of A and D is equal to the product of their individual probabilities, the two events are independent.

As for the event B and D, once again we can use either of the two conditions to check the independence of the two events. Note that B and D have one outcome in common which is (6,1). Therefore

$$P(B \cap D) = \frac{1}{36}$$

and

$$P(B) * P(D) = \frac{1}{6} * \frac{1}{6} = \frac{1}{36}$$

and since the two are equal, it shows that two events B and D the independent.

■ Example 20

In the 'Cancer and Smoking' example of the previous section, are the two events 'smoker' and 'die of cancer' independent? Clearly the answer is no, they are not independent since P(Cancer) = 0.24 and P(Cancer | Smoker) = 0.70.

▮ Example 21

It is known that 63% of the employees of a large company use the public transport to go to work. In a random sample of four employees, what is the probability that all four use public transport to go to work?

Let A_1 be the event that the first selected employee uses the public transport and define events A_2, A_3, and A_4 accordingly. Then

$$P(A_1) = 0.63, \; P(A_2) = 0.63, \; P(A_3) = 0.63 \text{ and } P(A_4) = 0.63$$

Now, since the four employees are randomly selected, the chances are that their decisions as to use or not use public transport to go to work are independent. Thus by the multiplicative rule for independent events, the probability that all four use the public transport to go to work is given by:

$$P(A_1 \cap A_2 \cap A_3 \cap A_4) = P(A_1).P(A_2).P(A_3).P(A_4) = 0.63 \times 0.63 \times 0.63 \times 0.63 = 0.63^4 = 0.1575$$

Exercise 14: A random sample of 400 patients at a clinic was taken and patients were classified according to two attributes: whether they were overweight or not overweight and whether they had high blood pressure or normal blood pressure. The following table summarizes the results:

Blood Pressure	Overweight	Not Overweight
High	154	75
Normal	48	123

Are the events 'being overweight' and 'having high blood pressure' independent? Why?

Exercise 15: In a quiz, there are eight 'True', 'False' questions. If you purely guess the answers, what is the probability that you get all questions right?

Random Variables

In the last chapter we learned the basic concepts of probability. We defined sample spaces and events. We saw that elements of a sample space may be symbols as in tossing a coin; they could be an ordered pair as in rolling a die twice, and so on. Here we are going to introduce a rule for associating a number to each outcome in the sample space. Such a rule is called a random variable. Random variables are very important in statistical analysis. They occur in various applications and are frequently used to define random phenomena in different experiments.

Section 1

A **random variable** is a rule that assigns a numerical value to each outcome of the sample space. Random variables are generally denoted by capital letters such as X or Y.

■ Example 1

Let X be defined as the number of tails in two tosses of a coin. We know that the sample space of this experiment has four outcomes and is given by

$$S = \{HH, HT, TH, TT\}$$

where, as before, H denotes the occurrence of the 'Heads' side of the coin and T denotes the outcome that the coin lands on its 'Tails' side. Clearly X, the number of 'Tails' in each outcome assigns a numerical value to each outcome in the sample space. Specifically, X assigns a value of 0 to the outcome of two 'heads,' i.e., HH, it assigns a value of 1 to each of the outcomes HT and TH and the value of X for the outcome TT is 2. Thus X is a random variable taking values 0 1, and 2.

■ Example 2

Let X be defined as the sum of the up faces when a die is rolled twice. As we learned in the previous chapter, the sample space of this experiment consists of 36 equally likely outcomes and each outcome consists of an ordered pair. Now, X is the sum of the pair of numbers for each outcome and again, X assigns a number to each outcome of the sample space. For example, X assigns a value of 4 to the outcome $(1, 3)$, a value of 7 to $(5, 2)$, a value of 10 to $(4, 6)$, and so on. In fact, X is a random variable taking the values $2, 3, \ldots, 12$.

■ Example 3

A coin is flipped until it lands on 'Tail'. Let X be defined as the number of flips in this experiment. Here, as we have seen before, the sample space of the experiment has infinitely many outcomes consisting of

$$S = \{T, HT, HHT, HHHT, HHHHT, \ldots\}$$

with X giving the value of 1 to the outcome T, value of 2 to HT, 3 to HHT, and so on. That is, the possible values of X are all positive integers $1, 2, 3, 4, 5, \ldots$.

■ Example 4

A number is picked at random from among all numbers between 0 and 1. Let X be the selected number.

Here, note that the nature of this example is different from the previous ones. In the last three examples, we were actually able to make a list of possible values of X. In this example, however, we cannot do so. We can see that if we take numbers in the interval of 0 to 1 to one decimal as in 0.0, 0.1, 0.2, $\ldots$, 1.0, then we are missing all numbers with two decimals. If we use a list of numbers with two decimals, we are missing all numbers with three decimals and so on. The fact is that in this case it is impossible to make a list of values of X and the set of values of X is what is called "uncountable." All we can say in order to express the set of values of X is to denote it as the closed interval of $[0, 1]$ or $\{x \mid 0 \leq x \leq 1\}$.

We are therefore going to distinguish between the random variable in this example and the random variables that we saw in examples 1, 2, 3. We call the random variables in examples 1, 2, 3, discrete and the random variable in example 4 a continuous random variable.

We call a random variable X **discrete** when the set of values of X is countable, i.e., we can make a list of those values. Otherwise if the set of values of X is uncountable, we call it **continuous**.

In this chapter, we deal only with discrete random variables and in the next chapter we discuss the continuous random variables. Therefore in this chapter, any time we refer to a random variable, we mean a discrete random variable.

Exercise 1

Determine whether the following random variables are discrete or continuous. Decide what you think the correct answer is AND provide a brief reason.

a. The body temperature of an individual

    Discrete   Continuous

b. The number of students in a class

    Discrete   Continuous

c. The number television sets in a household

    Discrete   Continuous

d. The amount of fluid in a soda pop

    Discrete   Continuous

e. The electricity consumption of a household in Pennsylvania

    Discrete   Continuous

    f. The time it takes to read a page in a book

        Discrete   Continuous

    g. The number of grammatical errors in manuscripts produced by an office

        Discrete   Continuous

Section 2: Probability Distribution

Now, consider the first example in this chapter, i.e., the example of flipping a coin twice and letting X be the number of Ts. We saw that X could take values of 0, 1, 2. Now, if we ask for the probability that X is 0, it is equivalent to the probability of the event $\{HH\}$ occurs and since the sample space consists of four equally likely outcomes, the probability of this event is $\frac{1}{4}$. Similarly, the probability that X is 1 is equivalent to the occurrence of the event $\{HT, TH\}$ which has a probability of $\frac{1}{2}$ and X being equal to 2 is the same as occurrence of the event $\{TT\}$ which has a probability of $\frac{1}{4}$. We can collect all this information in a table as

X	0	1	2
$P(X)$	1/4	1/2	1/4

Such a table is called the probability distribution of the random variable X. Note that we can also represent $P(x)$ graphically by drawing its histogram using probabilities as relative frequencies as in figure 4.1

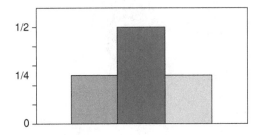

The **probability distribution,** $P(x)$, of a discreet random variable X is a function presented as a table or a formula that provides the probability for each value of X. Such a function has two properties:

i. Each value of $P(x)$ is a nonnegative number that is $P(x) \geq 0$ for every x.
ii. The sum of values of $P(x)$ over all values of X is equal to 1, i.e.,

$$\sum_x P(x) = 1$$

REMEMBER these two properties and be clear on what they mean!

■ Example 5

From past experience, a used car sales company knows that the distribution of the number of cars X that they sell in a week is as follows:

X	0	1	2	3	4	5
$P(X)$	0.05	0.15	0.25		0.10	0.10

Thus historically, the company knows that in 5% of the time they sold no cars in a week, 15% of the time they sold one car, and so on. They have never sold more than five cars in one week, but they are unsure about the proportion of the times that they sold three cars in one week. What is the probability that in a given week the company will sell:

a. Three cars?
b. At least two cars?
c. At most four cars?
d. Between two and four cars inclusive?

In order to answer the questions in this problem, we use the properties of a probability distribution. Specifically, in part a. because the sum of the probabilities of a probability distribution must be equal to one, in order to find the probability that the company sells three cars in a week, we find the sum of other probabilities in the table and subtract the total from one. Thus

$$P(3) = 1 - (.05 + .15 + .25 + .10) = 0.35$$

With the complete information in the table of probability distribution, we can easily calculate the probabilities regarding the other parts in this problem. Thus, for part b. we have

$$P(\text{at least 2 cars}) = P(X \geq 2) = .25 + .35 + .10 + .10 = 0.80$$

And for parts c. and d. we get

$$P(\text{at most 4 cars}) = P(X \leq 4) = 0.5 + .015 + .025 + .035 + .10 = 0.90$$

$$P(\text{between 2 and 4 cars}) = .25 + .35 + .10 = 0.85$$

■ Example 6

A bag contains three white and two red marbles. Two marbles are drawn at random and without replacement. If X is the number of red marbles drawn, find the probability distribution of X and sketch its graph.

Well, we have seen this problem several times in Chapter 3. In fact, we found that the probability of drawing one two white marbles is 0.3, the probability of drawing one red and one white marble is 0.6 and the probability of drawing two red marbles is 0.1. Hence, the probability distribution of X and its graph are as follows:

X	0	1	2
$P(X)$	0.3	0.6	0.1

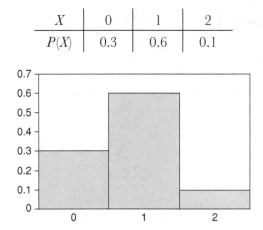

■ Example 7

In example 3, we defined the random variable X to be the number of times that a coin is flipped until it lands on its 'Tails' side. Here, we wish to derive the probability distribution of X. We note that in this case the set of values of X consists of all the positive integers. Therefore writing the distribution of X in a tabular format is clearly impractical. We therefore derive a mathematical formula that would provide the probability for any value of X. Specifically we try to look for a pattern of the probabilities and try to build the formula. Now we know that the probability that X is equal to one is the same the probability of a coin landing on its 'Heads' side in a single toss of the coin. Therefore

$$P(X = 1) = P(\{H\}) = 0.5$$

Also, since successive throws of a coin are independent events (as we saw in the last chapter,) we can say that

$$P(X = 2) = P(\{TH\}) = P(\{T\}) * P(\{H\}) = (0.5)^2$$

Similarly, we have

$$P(X = 3) = P(\{TTH\}) = P(\{T\}) * P(\{T\}) * P(\{H\}) = (0.5)^3$$

$$P(X = 4) = P(\{TTTH\}) = P(\{T\}) * P(\{T\}) * P(\{T\}) * P(\{H\}) = (0.5)^4$$

and so on. We can clearly see a pattern developing here. In fact, from the above we can conclude that the probability distribution of the random variable X can be described as

$$P(k) = P(X = k) = (0.5)^k \quad k = 1, 2, 3, \ldots.$$

From the above, we can obtain the probability for any value of X by simply substituting that value in the formula. How do we know that the sum of probabilities for all values of X from one to infinity is equal to 1? Well, mathematically we can actually prove that. However, the proof is outside the scope of this course and we will not present it here.

Exercise 2

The probability distribution of the number of courses taken by a randomly selected student X during a semester at a certain college is as follows:

X	1	2	3	4	5	6
$P(X)$	0.01	0.09	0.12		0.45	0.15

Find the probability and show ALL work that a student would take

a. exactly 4 courses

$P(4) =$

b. at least 3 courses

$P(3 \text{ or more courses}) =$

c. no more than 5 courses

$P(5 \text{ or less courses}) =$

d. between 2 and 5 courses inclusive

$P(2 \text{ to } 5 \text{ courses}) =$

Exercise 3

A bag contains six transistors, two of which are defective. Two transistors are drawn at random and without replacement from the bag. Find the probability distribution of the number of defective transistors in the sample.

Let X be the number of defective transistors in the sample then, the values of X are

$X =$

$$P(X = 0) = \text{............}$$
$$P(X = .\,) = \text{...........}$$

Therefore the probability distribution of X is:

X			
$P(X)$			

Exercise 4

If you roll a die until it lands on a '6,' what is the probability distribution of X, the number of times you roll the die? Remember to justify with an explanation!

Note that the possible values of X are all the positive integers namely, 1, 2, 3, 4 . . .

$P(X = 1) =$

$P(X = 2) =$

$P(X = 3) =$

$P(X = 4) =$

.

.

.

$P(X = k) =$ *for* $k = 1, 2, 3, 4, \ldots$

Section 3: Expected Value of a Random Variable

Suppose that your friend invites you to play a game. In this game, you roll a pair of dice and observe the sum of the number of pips on up sides of both dice. If this sum is 2, 3, 4, 5, 10, 11, or 12, you win $1. Otherwise if the sum is 6, 7, 8, 9, then you lose $1. Should you play this game?

Well, assuming that you play the game only if the odds are in your favor, the only way to answer this question is by considering the probabilities of winning and losing. At the first glance, it may seem that there are more numbers leading to winning each game, but let us look at the problem more carefully and consider the probability distribution of the random variable X defined as the sum of the up sides in a roll of a pair of dice. In Chapter 3, we learned that the sample space of the experiment of rolling a pair of balanced dice consists of 36 equally likely outcomes with probability of each sample point being equal to 1/36. By referring to that sample space, we can find the number of ways each integer from 2 to 12 can occur and find the respective probability of X being equal to that number.

For example, the probability that X is equal to 2 is 1/36 because there is only one outcome, namely (1,1), i.e., a double one that leads to a total of 2. Similarly, the probability that X is equal to 3 is 2/36 because there are two outcomes in the sample space (1,2) and (2,1) giving rise to a total of 3. The probability that X is 4 is 3/36 since the three outcomes (1,3), (2,2), and (3,1) gives a total of 4. Continuing in this manner, we find the following probability distribution for X:

X	2	3	4	5	6	7	8	9	10	11	12
$P(X)$	1/36	2/36	3/36	4/36	5/36	6/36	5/36	4/36	3/36	2/36	1/36

It is easy to verify that the sum of the probabilities in the above distribution is indeed equal to 1 as it should be. Using this distribution, we can calculate the probability of winning and losing the game. Remember you win the game if the sum of the up sides is 2, 3, 4, 5, 10, 11, and 12. Thus adding the corresponding the probabilities,

$$P(\text{win}) = \frac{1}{36} + \frac{2}{36} + \frac{3}{36} + \frac{4}{36} + \frac{3}{36} + \frac{2}{36} + \frac{1}{36} = \frac{16}{36} = \frac{4}{9}$$

Similarly, we can calculate the probability of losing by adding the probabilities of the sum being 6, 7, 8, or 9, that is

$$P(\text{lose}) = \frac{5}{36} + \frac{6}{36} + \frac{5}{36} + \frac{4}{36} = \frac{20}{36} = \frac{5}{9}$$

We can now clearly see that this game is not a fair game and your chance of losing is higher than your chance of winning. In fact if we define a random variable Y as your gain in the game, the probability distribution of Y can be expressed as follows:

Y	−1	+1
$P(Y)$	5/9	4/9

This means that if you keep on playing this game a large number of times, on average out of every nine games you expect to lose five and win four. Thus your average expected gain per game can be calculated as

$$(-1) * 5/9 + (+1) * 4/9 = (-5 + 4)/9 = -1/9 = -0.1111$$

This is what we call the expected value of the random variable Y. What it means is that if you play the game for a large number of times, on average you expect to lose $0.1111 or just over a dime per game. Note that we are not saying that if you play the game once, you lose a dime. That statement, of course, does not make any sense in this case. Clearly, if you play the game only once, you either lose a dollar or win a dollar. But, in a large number of replications of the game you can expect to lose an average of $0.1111 per game. The mean of a random variable is shown by μ which is the same as its expected value $E(Y)$. Thus in this example,

$$\mu = E(Y) = -0.1111$$

If X is a random variable with probability distribution $P(X)$, then the **mean** or the **expected value** of that random variable, being the long term average of that random variable, is defined as

$$\mu = E(X) = \sum X * P(X)$$

Therefore to compute the mean of a random variable, we multiply each value of the random variable by its respective probability and add the results.

■ Example 8

In example 5, we discussed the distribution of the number of cars that a used car sales company sells in a given week. Here, we are interested to know how many cars the sales company can expect to sell in a week. We have

X	0	1	2	3	4	5
$P(X)$	0.05	0.15	0.25	0.35	0.10	0.10
$X * P(X)$	0	0.15	0.50	1.05	0.40	0.50

Thus

$$\mu = E(X) = 0.15 + 0.50 + 1.05 + 0.40 + 0.50 = 2.60 \text{ cars}$$

which means that the sales company can expect to sell on average 2.6 cars per week. Of course this, again, does not say that in any given week the sales company sells 2.6 cars. That clearly is a meaningless statement since the number of cars sold per week has to be an integer. But it only gives the average number of cars they sell over a long period of time.

■ Example 9

In example 6, we found the distribution of the number of red marbles drawn when two marbles are drawn at random from a bag that contains three white and two red marbles. Here, we find the mean of that distribution. We have,

X	0	1	2
$P(X)$	0.3	0.6	0.1
$X * P(X)$	0	0.6	0.2

From which we get

$$\mu = E(X) = 0.6 + 0.2 = 0.8 \text{ red marble}$$

Once again we should emphasize that this does not mean that when we draw two marbles out the bag with three white and two red marbles then we draw 0.8 of a red marble and 1.2 white marbles. That clearly is a meaningless statement. Rather, what it means is that if we draw two marbles out of the bag and note the number of red marbles and repeat this process a large number of times, then the average number of red marbles per draw is 0.8.

Exercise 5

Find the mean number of courses taken by students when the distribution of the number of courses is given by (as in the last section)

X	1	2	3	4	5	6
$P(X)$	.01	.09	.12	.18	.45	.15
$X * P(X)$						

$$\mu = E(X) = \ldots\ldots$$

Section 4: Variance and Standard Deviation

When we were studying the properties of a sample in Chapter 2, we learned that the mean alone does not provide sufficient information to characterize the sample since the mean or the median for that matter only give us information about the center of the data set and give us no information about the spread and variation of the data points. By the same argument, the mean of a random variable does not provide any information about the variation in the values of that random variable and we need to have a measure of dispersion to characterize the spread in the values of the random variable. We also recall that the variance of a sample was calculated based on sum of the squares of deviation. Similarly we define the variance of a random variable.

To find the **variance** of a random variable X denoted by σ^2, we first subtract the mean μ from each value of X to find the deviations. We then square the deviation, multiply by its respective probability and add. That is,

$$\sigma^2 = E(X - \mu)^2 = \sum (X - \mu)^2 P(X)$$

Similar to what we learned in Chapter 2, the unit for measurement of variance is the square of the units with which the values of the random variable are measured. In order to have a measure of variability which has the same unit as the values of the random variable, we define the standard deviation σ of a random variable as the square root of the variance.

■ Example 10

For the distribution of the number of cars sold in a week by the sales company described in Example 5, we found that the mean was 2.6 cars. Here, we wish to compute the variance and standard deviation of the number of cars sold per week. We also want to compute the probability that the number of cars sold in a week is within one standard deviation of its mean.

From Example 6 of this chapter, we had

X	0	1	2	3	4	5
$P(X)$	0.05	0.15	0.25	0.35	0.10	0.10
$X * P(X)$	0	0.15	0.50	1.05	0.40	0.50
$(X - \mu)^2 * P(X)$	0.338	0.384	0.09	0.056	0.196	0.576

where the numbers in the bottom row are found by first subtracting the mean $\mu = 2.6$ from the value of X, squaring it and multiplying by the probability for that value. The variance is now found by adding the numbers in the bottom row. Thus

$$\sigma^2 = 0.338 + 0.384 + 0.09 + 0.056 + 0.196 + 0.576 = 1.64 \ (\text{cars})^2$$

and, as before, once we have the variance, we can find the standard deviation by simply taking its square root, i.e.,

$$\sigma = \sqrt{1.64} = 1.28 \text{ cars}$$

Now, if we are interested in finding the probability that the number of cars that they sell in a week is say within one standard deviation of its mean, we have

$\mu - 1\sigma = 2.6 - 1.28 = 1.32$
$\mu + 1\sigma = 2.6 + 1.28 = 3.88$

and therefore

$$P(\mu - \sigma \leq X \leq \mu + \sigma) = P(1.32 \leq X \leq 3.88) = P(X = 2, 3) = 0.60$$

It is to be noted that similar to what we discussed in Chapter 2, we can use the Chebyshev's rule and the empirical rule to interpret the standard deviation. In fact, according to the Chebyshev's rule, we can say that for any random variable X, irrespective of the shape of its distribution, the probability that X takes a value within k standard deviations of its mean, where k is any real number, is at least $1 - 1/k^2$. Also, if the shape of the distribution is the normal curve, the empirical rule will apply. We shall see more about the normal curve in the next chapter.

YOUR TURN! Please complete the required exercises below directly in this book. You are encouraged to discuss each exercise with a partner or group. You may be asked complete the exercises in class or outside of class.

Exercise 6

Find the variance and standard deviation of the number of courses taken by students when the distribution of the number of courses is given by (as in the last section)

X	1	2	3	4	5	6
$P(X)$	0.01	0.09	0.12	0.18	0.45	0.15
$(X-\mu)^2 * P(X)$						

$\text{Var}(X) = \sigma^2 = \dots\dots$ $\sigma = \dots$

Compute also the probability the number of courses taken by a randomly selected student is within one standard deviation of its mean

$\mu - \sigma = \dots\dots$

$\mu + \sigma = \dots\dots$

$P(\mu - \sigma \le X \le \mu + \sigma) = \dots\dots$

Section 5: Binomial Distribution

In our daily life, there are many instances that we repeat a process and we are interested in the number of occurrences of a certain outcome. For example, you might ask ten fellow students whether they think you should take a certain course or not and you are interested in the number out of ten who would say yes. I might ask a number of residents in my town whether or not they agree with the new town ordinance and would be interested in the number of those who agree. A surgeon might be interested in the number of individuals out of 15 heart transplant patients who have survived ten years after their surgery. A pollster is interested in the number of individuals from a random sample of say 25 who are going to vote for a certain candidate and so on. The fact is that all of these experiments fall under the umbrella of binomial experiments. We define a binomial experiment as an experiment with the following four characteristics:

- It consists of a fixed number n of identical trials.
- Each trial can have one of two possible outcomes, which we name 'Success' and 'Failure.'
- Successive trials are independent.
- Probability of success per trial p is the same for every trial.

The binomial random variable X is defined as the number of successes out of the n trials. Clearly, X can assume any integer from zero to n (the number of trials) inclusive. Our goal in this section is to derive a general formula for the distribution of the binomial random variable. In order to do so, once again, we will use an example and try to find patterns that lead us to the formula for the binomial distribution.

■ Example 11

In a large city, it is known that 70% of people are in favor of building a new sports complex for the city. In a random sample of four citizens, if X is the number of people who are in favor of this project, we wish to find the probability distribution of X.

We first note that the experiment in this problem satisfies all the four conditions of the binomial distribution stated above with $n = 4$ and $p = 0.7$. Therefore X is a binomial random variable. If we denote by S the outcome of a person in favor and by F the outcome of a person against the project, then the sample space of this experiment has 16 outcomes and can be described as follows:

$$S = \{FFFF \quad FFFS \quad FFSS \quad FSSS \quad SSSS$$
$$FFSF \quad FSFS \quad SFSS$$
$$FSFF \quad FSSF \quad SSFS$$
$$SFFF \quad SFSF \quad SSSF$$
$$SSFF$$
$$SFFS\}$$

We note that these outcomes are not equally likely since the probability of each 'S' is 0.7 and the probability of each 'F' is 0.3. Imagine having a coin that is loaded in such a way that it lands on 'Head' side 70% of the times and on 'Tails' side only 30% of the times. If you flip this coin in the air four times, clearly the chance that this coin lands on its 'Heads' side all four times is not the same that it lands on its 'Tails' side and is higher. Therefore to think that each of the outcomes in the above sample space has a probability of 1/16 is erroneous. Note also that the way the sample space is organized, the random variable X, the number of successes out of four trials, assigns a value of 0 to the outcome in the first column, a value of 1 to each of the outcomes in the second column, a value of 2 to each of the outcomes in the third column, and so on. Using the fact that successive trials in a binomial experiment are independent, we have

$$P(0) = P(X = 0) = P(FFFF) = P(F)*P(F)*P(F)*P(F) = (0.3)^4$$

Similarly, to find $P(X = 1)$, we note firstly that there are four combinations of outcomes in the sample space that lead to the value of X being equal to 1. Each of these combinations consists of one success and three failures. We therefore have

$$P(1) = P(X = 2) = 4\ P(SFFF) = 4\ (0.7).(0.3)^3$$

Continuing in this manner, we get the following probability distribution for X

X	0	1	2	3	4
$P(x)$	$(0.3)^4$	$4\ (0.7)\ (0.3)^3$	$6\ (0.7)^2\ (0.3)^2$	$4\ (0.7)^3\ (0.3)$	$(0.7)^4$

We purposely left the probabilities in the above table uncalculated since we are looking for a pattern to determine the probability distribution of a binomial random variable in general. But, if we calculate the probabilities, then it is quite easy to see that the probabilities sum to one as they should.

Now, once again note that in the above example the number of trials n is only four and still we had a fairly large sample space. Clearly, as the number of trials increases, the size of the sample space also grows. In fact, for each additional trial, the size of the sample space grows by a factor of two. Thus for five trials, we would have a sample space with 32 outcomes, with six trials the sample space would have 64 trials, and so on. Thus writing the sample space in each case for each experiment to find the probabilities is not feasible. For example, for a binomial experiment with 15 trials, the sample space would consist of $2^{15} = 32,768$ outcomes. However, if we look at the table of the probability distribution of the random variable X in the above example, we see that each probability consists of three factors. One factor is the probability of success raised to the number of successes; another factor is the number of failures raised to the number of failures and yet there is another factor which is the number of combinations of the number of successes out of the four trials. Hence, we can conclude that in general if there are n trials in a binomial experiment then the probability of x successes would consist of a factor p^x which is probability of success raised to the number of successes, a factor $(1 - p)^{n-x}$ which is the probability of failure raised to the number of failures. We also need a factor that would give us the number of combinations of x successes out of n trials. Well, fortunately in mathematics there is a formula that gives us the number of different selections of x items out of n items. This number, denoted by $\binom{n}{x}$ is given by

$$\binom{n}{x} = \frac{n!}{x!(n-x)!}$$

where $n!$ is called the factorial of the number n and it is the product of the integer n and all of its predecessors down to 1, for example,

$$6! = 6 \times 5 \times 4 \times 3 \times 2 \times 1 = 720$$

Most scientific calculators have a key for the factorial of a number. This key is generally in the mathematical functions, under the selection of probability functions. Moreover, scientific calculators also have a key to give the value of $\binom{n}{x}$ for given n and x. Often the function is denoted by C_x^n where the letter C stands for 'combination' since this is the number of combinations of x items out of n items. Again, the function is generally found under probability functions on the calculator.

■ Example 12

Let us, for practice, determine the values of

a. $14!$ b. $14!/6!$ c. $\dfrac{14!}{6!8!}$ d. $\begin{pmatrix} 14 \\ 6 \end{pmatrix}$ e. $\begin{pmatrix} 14 \\ 8 \end{pmatrix}$

Using a scientific calculator, we find

a. $14! = 87{,}178{,}291{,}200$

b. $14! / 6! = 121{,}080{,}960$

c. $\dfrac{14!}{6!8!} = 3003$

d. $\begin{pmatrix} 14 \\ 6 \end{pmatrix} = \dfrac{14!}{6!(14-6)!} = 3003$

e. $\begin{pmatrix} 14 \\ 8 \end{pmatrix} = \dfrac{14!}{8!(14-8)!} = 3003$

Note that the answers to parts d. and e. are the same. The reason is that the number of different selections of 6 items out 14 is the same the number of combinations of 8 items out of 14. If I have 6 apples and 8 bananas and wish to divide the fruits among 14 children in daycare center, I can either choose 6 children first and give them an apple each and give the bananas to the remaining 8 or choose 8 children first to give them a banana each and give the apples to the remaining 6 children. The number of selections is the same in both cases.

Exercise 7

Compute the following using a calculator.

a. 11!

b. 11!/4!

For c and d, first write out what the notation means, then use a calculator to find your final answer.

c. $\begin{pmatrix} 11 \\ 4 \end{pmatrix}$

d. $\begin{pmatrix} 11 \\ 7 \end{pmatrix}$

Exercise 8

From a class of 15, we wish to select 5 for a certain project. In how many different ways can we make this selection? SHOW ALL WORK.

Probability Distribution of Binomial Random Variable

Returning now to our derivation of the formula for the binomial distribution, we can finally say that the probability distribution of the binomial random variable X is given by

$$P(x) = \binom{n}{x} p^x (1-p)^{n-x} \quad x = 0, 1, 2, \ldots, n$$

Thus given the values of n and p, we can determine the probability of x successes by using the above formula. Many scientific calculators have a special key for the binomial distribution. In the Texas Instrument calculators, this function is listed under the probability distributions and is called 'binompdf.' The function requires three arguments n, p, and x in that order. So, for example if 32% of the population of a city live in the suburbs, then in a sample of 26 people, the probability the number out of 26 who live in the suburbs is say 12 can be found by binompdf (26, 0.32, 12) = 0.0503.

■ Example 13

In example 11 of this chapter, we considered a problem where a sample of four individuals was taken from a city in which 70% of residents are in favor of building a new sports complex for the city and we were interested in the distribution of the number of individuals out of four who were in favor of the project.

QUESTION: Suppose now that instead of four individuals, more realistically, we take a sample of 20 people from the city. Then we would like to calculate the probability that the number of people who are in favor in the sample of 20 is:

a. 17
b. 12
c. At most 17
d. At least 12
e. Between 12 and 17 inclusive

SOLUTION: Let X be the random variable designating the number of individuals out of 20 who are in favor of the project of building a new sports complex for the city. Then, clearly X has a binomial distribution with $n = 20$ and $p = 0.7$. Therefore, we have

a. $P(17) = \binom{20}{17}(0.7)^{17}(1 - 0.7)^{20-17} = \binom{20}{17}(0.7)^{17}(0.3)^3 = 0.0716$
b. $P(12) = \binom{20}{12}(0.7)^{12}(1 - 0.7)^{20-12} = 0.1144$

Before we proceed to part c., it is worth mentioning that both parts a. and b. could have been solved using the calculator. For a., we have

$$P(17) = \text{binompdf} \ (20, 0.7, 17) = 0.0716$$

And for part b.

$$P(\ 12) = \text{binompdf} \ (20, 0.7, 12) = 0.1144$$

Now,

c. $P(X \leq 17) = P(0) + P(1) + P(2) + \ldots + P(17)$

Thus to find the probability that at most 17 people are in favor, we need to use the binomial formula to calculate the probabilities of 0, 1, . . ., 17 and add the results. This process can be very time consuming, but at the end of almost all statistical textbooks, one can find a table of cumulative probabilities of the binomial distribution. In addition, some scientific calculators provide this cumulative probability distribution function as well among the statistical distributions that the calculator carries. In the Texas Instrument calculators, this function is 'binomcdf.' Similar to binompdf, this function also requires the three arguments n, p, x in that order, but binomcdf provides the cumulative sum of probabilities. Thus the distinction between these functions on the calculator is quite crucial. The function binompdf (n, p, x) gives the probability that the binomial variable X assumes the value x, whereas binomcdf (n, p, x) gives the probability that X assumes any of the number s from 0 to x which is the sum of probabilities from 0 to x. Using either the binomial tables or a calculator, we find

$$P(X \leq 17) \ \text{binomcdf} \ (20, 0.7, 17) = 0.9645$$

d. Here, we want $P(X \geq 12)$. Thus our one option is to use the binomial formula and to compute $P(12), P(13), \ldots, P(20)$. But, a better and shorter way to find the probability that at least 12 people are in favor of the project is to use the law of complements and utilize the binomial tables or the cumulative function key for the binomial distribution on a calculator. We have

$$P(X \geq 12) = 1 - P(X \leq 11) = 1 - \text{binomcdf} \ (20, 0.7, 11) = 1 - 0.1133 = 0.8867$$

e. Similarly,
$P(12 \leq X \leq 17) = P(X \leq 17) - P(X \leq 11) = \text{binomcdf} \ (20, 0.7, 17) - \text{binomcdf} \ (20, 0.7, 11) = 0.9645 - 0.1133 = 0.8512.$

Exercise 9

It is known that 60% of a certain plant's seeds germinate. If we plant 15 seeds, find the probability that number of seeds that germinate is:

a. Exactly 12? Show your work

$P(12) =$

b. Exactly 5? Show your work

$P(5) =$

c. At most 12? Show your work

$P(X \leq 12) =$

d. At least 5?

$P(X \geq 5) =$

e. Between 5 and 12 inclusive?

$P(5 \leq X \leq 12) =$

Section 6: Mean and Variance of Binomial Distribution

For the binomial random variable X, how can we find the mean value of the variable and how can we calculate the variance and standard deviation? Well, as we learned in the previous section, in order to compute the mean of a discrete random variable, we need to multiply each value of that random variable by its respective probability and add the results. Therefore in order to find the mean of a binomial random variable X it appears that we have to first compute the probability for each value of X, multiply this by the value of X and add. Similarly, we saw in the last section that in order to compute the variance and hence the standard deviation of a discrete random variable, we need to first calculate the square of the deviation for each value of the random variable, multiply it by the respective probability and add the results. This seems like a significant amount of work for the binomial random variable, especially if the number of trials n is rather large. But, fortunately, there are simple mathematical formulas that actually produce the mean, variance, and standard deviation of a binomial random variable. These formulas are based on strict mathematical derivations that use the definitions of mean and variance in the general form. Description of those derivations is outside the scope of this course and here we just state these formulas and use them in examples. If n is the number of trials and p is the probability of success per trial, then the mean and variance of the binomial random variable X, that is the number of successes in n trials are given by:

$$\mu = n * p$$

$$\sigma^2 = n * p * (1-p)$$

These two formulas quite simply provide the mean and variance and clearly, once we have the variance, we can find the standard deviation by taking the square root of the variance.

■ Example 14

In the previous example we had a binomial distribution in which n was 20 and p was 0.7. That is, we were interested in the number of individuals out of 20, who favored building a new sports complex for the city when we knew that 70% were in favor of the project. Now, in our sample of 20, how many, on average, can we expect to be in favor? What is the variation, as measured by the standard deviation, in the number out of 20 who are in favor of the project? And what is the probability that the number in favor out of 20 is within one standard deviation of the mean?

Well to answer these questions, we use the above formulas. We have

$$\mu = 20 * 0.7 = 14$$

which means that on average we can expect to have about 14 in our sample to be in favor. This makes sense as 70% of 20 is 14. Also, the variance is given by

$$\sigma^2 = 20 * 0.7 * (1-0.7) = 4.2$$

and therefore

$$\sigma = \sqrt{4.2} = 2.049$$

Now, in order to compute the probability that the number of individuals in favor of the project in a sample of 20 is within one standard deviation of the mean, as before, we have

$$\mu - \sigma = 14 - 2.049 = 11.951$$

$$\mu + \sigma = 14 + 2.049 = 16.049$$

Hence

$$P(\mu - \sigma \leq X \leq \mu + \sigma) = P(11.951 \leq X \leq 16.049) = P(X = 12, 13, 14, 15, 16)$$

$$= P(X \leq 16) - P(X \leq 11) = \text{binomcdf } (20, 0.7, 16) - \text{binomcdf } (20, 0.7, 11) = 0.7796$$

In Chapter 2, we learned that according to the empirical rule, about 68% of values in the distribution are within one standard deviation of the mean. Why does the above value not conform to the empirical rule and give us a value close to 0.68? Well, we must remember that the empirical rule only applies to the normal curve.

Exercise 10

Continuing in the last exercise, if 60% of the seeds of a plant actually germinate when planted, in a random sample of 15 seeds, what is

a. The expected number of seeds that germinate? Show each step you take.

b. The standard deviation of the number of seeds that germinate? Show each step you take.

c. The probability that the number of seeds that germinate is within one standard deviation of its mean? Show each step you take.

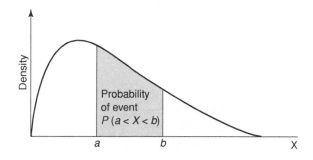

Chapter 5

Continuous Random Variable

Section 1: Introduction

At the beginning of the last chapter we learned that there are two types of random variables: Discrete and Continuous. We discussed the properties of discrete random variables in Chapter 4. Here, we consider the properties of continuous random variables. Recall that a continuous random variable is that whose set of possible values is not countable, i.e., its values cannot be listed. For example, measuring the height of an individual or measuring the weight of a baby at birth. These random variables can take any real value within a given range and it is impossible to make a list of the possible values. Therefore similar to what we saw in the last chapter, if we wish to draw a histogram for the distribution of a continuous random variable, then histogram consists of infinitely many rectangles. For example, suppose we wish to draw a histogram for the distribution of the weight of infants at birth. If we measure these weights to the nearest pound, we can draw a histogram with rectangles most likely based on say 3, 4, 5, 6, 7, 8, 9, and 10. Now, suppose the weights are measured nearest to the tenth of a pound. Then the number of rectangles becomes 10-fold. If we measure the weights to the hundredth of a pound, the number of rectangles will be 100-fold and so on. In the limit, the histogram will consist of infinitely many rectangles and therefore becomes a smooth curve. This smooth curve is called the **density curve** of a random variable. Thus associated with any continuous random variable X, we define a density function $f(x)$ with the following two properties:

- Graph of $f(x)$ is entirely above x-axis, i.e., $f(x) \geq 0$ for every X.
- Total area under density curve is equal to 1.

The density curve characterizes the random variable and displays its distribution. For continuous random variables, the probability that the variable assumes a value between two given numbers such as a and b, i.e., $P(a < X < b)$ is the area under density curve between a and b. Note also

that because of continuity, probability that the variable X takes any specific value is 0, i.e., $P(X = c) = 0$ for any c. Thus for a continuous random variable, we have

$$P(a < X < b) = P(a \leq X < b) = P(a < X \leq b) = P(a \leq X \leq b)$$

■ Example 1: Bus Delay

QUESTION: A bus is scheduled to make a stop at a certain bus station on the hour every hour from 7 am to 11 pm. However, because of late afternoon traffic, early evening buses are often late by as much as 10 minutes, with the same chance of arrival at any instant of time in that 10-minute interval. Find the probability that the 6 pm bus is late by

a. at most 4 minutes
b. exactly 4 minutes
c. less than 4 minutes
d. more than 7 minutes
e. between 3 and 8 minutes inclusive

Let X denote the amount of time by which the bus is late. Then clearly X is a continuous random variable and $0 \le X \le 10$. Moreover, because the bus is equally likely to arrive at any instant within the 10-minute interval, we can expect the graph of the density of X to be a flat curve assigning the same weight to every point. Thus the density curve will look like a rectangle with length 10

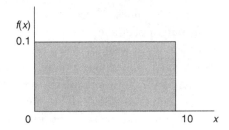

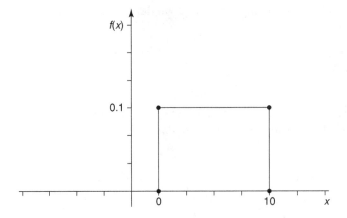

and because the total area under the curve must be equal to 1 and the area of a rectangle is length times width, the width of the density curve must be 0.1 so that

$$Area = 10 \times 0.1 = 10 \times \frac{1}{10} = 1$$

Now, recall that the probability that a continuous random variable takes a value between two given numbers is represented by the area under the density curve between the two numbers. Therefore in part a, to find the probability that the bus is late by at most 4 minutes, we are seeking the probability that X takes a value between 0 and 4. Hence, that probability is equal to the area of the rectangle between 0 and 4 and thus,

$$a.\ P(0 \le X \le 4) = 4 \times 0.1 = 4 \times \frac{1}{10} = 0.4$$

As discussed before, another property of continuous random variables is that because of continuity the probability that the variable takes a specific value is 0. Thus in part b, the probability that the bus is delayed by exactly 4 minutes is 0 and we have

b. $P(x = 4) = 0$

And as a consequence of part b, we can readily see that part c is effectively asking the same question as part a and therefore

c. $P(X < 4) = .4$

We can now easily see that parts d and e are respectively the area under the rectangle between 7 and 10, and the area under the rectangle between 3 and 8. Thus,

d. $P(7 < X < 10) = P(x > 7) = 3 \times 0.1 = 0.3$

e. $P(3 \leq X \leq 8) = 5 \times 0.1 = 0.5$

The above example is a case of a more general type of distributions. When the continuous random variable is evenly distributed throughout a continuous range, we say that the variable has a uniform distribution.

Section 2: Uniform Distribution

A random variable is said to have a uniform distribution over the interval (c, d) when its density function is given by:

$$f(x) = \frac{1}{d - c}$$
$$c < X < d$$

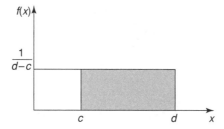

and the graph of $f(x)$ is a rectangle for which the length is $d - c$ and the width is $\frac{1}{d - c}$. For a uniform random variable, the mean and standard deviation are given by:

$$\mu = \frac{d + c}{2} \qquad \sigma = \frac{d - c}{\sqrt{12}}$$

where once again, we mention that the above formulas are based on rigorous mathematical derivation; however, description of the derivations is outside the scope of this course.

■ Example 2: Bus delay

In the last example, find mean delay of the bus. Find also standard deviation and compute the probability that the bus is delayed by a time which is within one standard deviation of its mean.

Note that in the bus delay example, $c = 0$ and $d = 10$ and therefore,

$$f(x) = \frac{1}{10} \qquad 0 < X < 10$$

$$\mu = \frac{10+0}{2} = 5 \qquad \sigma = \frac{10-0}{\sqrt{12}} = 2.88$$

as in the previous chapter to find the probability that X takes a value within one standard deviation of the mean, we have

$$\mu - \sigma = 5 - 2.88 = 2.12$$
$$\mu + \sigma = 5 + 2.88 = 7.88$$

The difference, however, is that in the present situation we find the probability of using the area under the curve, that is

$$p(\mu - \sigma \leq X \leq \mu + \sigma) = P(2.12 < X < 7.88) = (7.88 - 2.12) \times 0.1 = .576$$

Exercise 1

The time that it takes an office secretary to respond to an email inquiry in minutes, has a uniform distribution between 1 and 19. Draw the graph of the density function and

a. Find the probability that an email would take longer than 15 minutes.
b. What proportion of emails would take less than 10 minutes?
c. What is the probability that an email would take between 5 and 7 minutes?
d. What is the average time for an email?
e. What is the standard deviation of the time it takes to respond to an email?
f. What proportion of email responses are within one standard deviation of the average?

In this problem,

$$c = \text{_____} \quad \text{and} \quad d = \text{_____}$$

Therefore

$$f(x) = \text{_____}$$

and the graph of $f(x)$ is

Now,

a. $p(X \geq 15) = P(15 \leq X \leq 19) = $ _____

b. $p(X \leq 10) = $ _____

c. $P(5 \leq X \leq 7) = $ _____

d. $\mu = $ _____

e. $\sigma = $ _____

f. $P(\mu - \sigma \leq X \leq \mu + \sigma) = $ _____

Section 3: Normal Distribution

In this section, we introduce one of the most important statistical distributions, namely the normal distribution. Formally, we say that a random variable X has a normal distribution when its density is the normal curve, in which case we call X a normal random variable. In Chapter 2, we briefly learned some properties of the normal curve. We learned that the normal curve is symmetric, unimodal, and bell-shaped. The axis of symmetry is at the median of the distribution, which also coincides with the mean. Moreover, we learned that the empirical rule only applies to the normal curve. It turns out that the normal distribution is actually characterized by two quantities:

1. The mean μ—which is the center of the distribution
2. Standard deviation σ which determines the spread of the normal curve

Therefore, once we know the above two quantities, we know where the distribution is centered and how wide or narrow the curve is. Note that in any case the total area under the curve is equal to one. In fact the mathematical formula for the density of normal curve is given by:

$$f(x) = \frac{1}{\sigma\sqrt{2\pi}} e^{\frac{(x-\mu)^2}{2\sigma^2}} \qquad -\infty \leq x \leq \infty$$

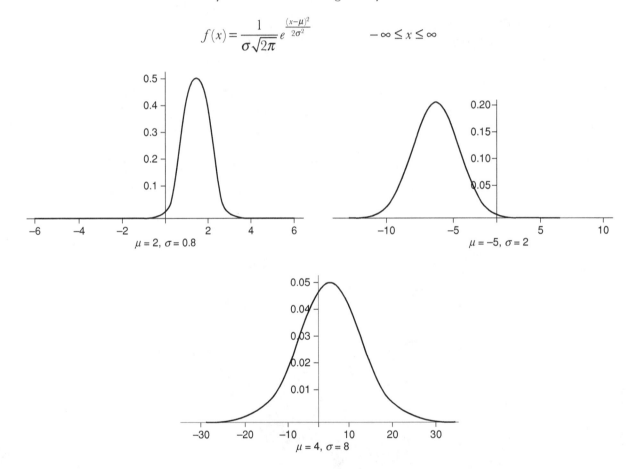

Therefore if we use our graphing calculator and graph the above function, using some arbitrary values for μ and σ, then we will see the graph of the normal curve. Now, because the normal random variable is a continuous random variable, as we discussed earlier, the probability that the normal variable X assumes a value between two numbers a and b, i.e., $P(a < X < b)$ is given by the area under the corresponding normal curve.

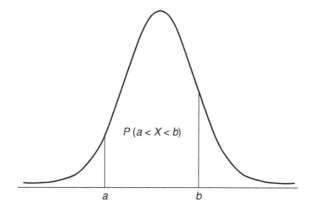

$P(a < X < b)$

a   b

So the question is how we can find the area under the normal curve between two given numbers? In the previous section, where we discussed the uniform distribution, finding the area under the curve was not a problem as every time we had a rectangle and finding the area of a rectangle is straightforward. The difficulty here is that the normal curve does not resemble any of the geometric shapes that we have seen before. Our goal in this section is to learn how we can handle the problem of finding the area under the normal curve and consequently the probabilities relating to a normal random variable.

We begin by considering a special case. Among all the infinitely many ways that we can draw a normal curve, we are going to pick one for which the center of the curve, i.e., the mean of the distribution is at zero and the measure of the spread, that is the standard deviation is equal to one,

$$\mu = 0 \qquad \sigma = 1$$

We call this special normal distribution the **standard normal**, and use Z to show the corresponding random variable. For this case, area under the curve is available in tabular form. At the back of almost any statistics textbook, there is a chart that provides the area under the curve. In the final pages of this manuscript, Table 1 provides a copy of the normal distribution table that gives the area between the center 0 and any given value. Note in the first column of that the table does not go beyond 3.0 simply because by the empirical rule we know that 99.7% of Z values in a normal distribution are between −3 and +3.

■ Example 3: Area Under the Standard Normal Curve Using Normal Tables

$P(0 < Z < 2.16) = 0.4846$

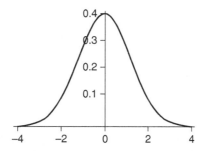

We can also find this chart on the Internet. Moreover, most of the modern graphing calculators have a special key that provides the area under the standard normal. In the Texas Instrument calculators (TI 84, 85), this function is called *normalcdf*. Thus if, for example, we want to find the probability that the standard normal variable Z takes a value say between 1 and 2, i.e., the area under the standard normal curve between 1 and 2, the function *normalcdf* (1,2) will provide the answer.

■ Example 4: Calculating Probability for Standard Normal Variable

QUESTION: Find the following for a standard normal random variable Z

a. $P(0 < Z < 1.82)$

b. $P(Z < -1.3)$

c. $P(Z = -1.35)$

d. $P(Z \leq -1.35)$

e. $P(Z > 2.43)$

f. $P(-1.88 < Z < 0.75)$

g. $P(-1 < Z < 1)$

h. $P(-2 < Z < 2)$

i. $P(-3 < Z < 3)$

j. a constant c such that $P(Z < c) = 0.9834$

k. 90th percentile of distribution

SOLUTION (Part a): Using the standard normal chart, we find

$$P(0 < Z < 1.82) = 0.4656$$

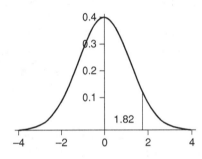

We could also find this probability using the calculator,

$$P(0 < Z < 1.82) = normacdf\,(0,1.82) = 0.4656$$

SOLUTION (Part b):

$$P(Z < -1.35) = 0.5 - 0.4115 = 0.0885$$

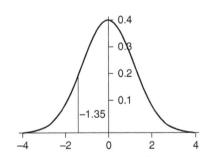

Now, in order to use the calculator to find this area, we must remember that the function on the calculator provides the area between two numbers and here we are seeking the entire area to the left of -1.35. However, we note that the entire area to the left of -1.35 means the area between $-\infty$ and -1.35. Some of the more recent TI calculators have a key for infinity but in the older versions, the calculator does have a key for $-\infty$ or $+\infty$; in practice, we use a very small number or a very large number to represent them. For example, for $-\infty$, we use 10^{-99} and for ∞, we use 10^{99}. Thus,

$$P(Z < -1.35) = normalcdf\,(-8\,,-1.35) = 0.0885$$

SOLUTION (Part c): Clearly since Z is a continuous random variable, the answer to this part of the example is trivial and we have

$$P(Z = -1.35) = 0$$

SOLUTION (Part d): In light of part c we immediately see that

$$P(Z \leq -1.35) = 0.0885$$

SOLUTION (Part e):

$$P(Z > 2.43) = 0.5 - 0.4925 = 0.0075$$

Equivalently,

$$P(Z > 2.43) = 0.5 - normalcdf(0, 2.43) = normalcdf(2.43, \infty) = 0.0075$$

SOLUTION (Part f):

$$P(-1.88 < Z < 0.75) = 0.2734 + 0.4699 =$$

Also,

$$P(-1.88 < Z < 0.75) = normalcdf(-1.88, 0.75) =$$

SOLUTION (Part g):

$$P(-1 < Z < 1) = 0.3413 + 0.3413 = 0.6826$$

Also,

$$P(-1 < Z < 1) = normalcdf(-1, 1) = 0.6826$$

SOLUTION (Part h):

$$P(-2 < Z < 2) = normalcdf(-2, \ 2) = 0.9544$$

SOLUTION (Part i):

$$P(-3 < Z < 3) = normalcdf(-3, 3) = 0.9974$$

NOTE: that parts g, h, and i constitute the empirical rule that we discussed in Chapter 2.

SOLUTION (Part j): Now, in this part of the problem, we see that the nature of the question is different from the other parts. In the previous parts of this problem, some Z values were given and calculation of probability or area under the curve was required. Here, the probability or area is given, and the corresponding Z value is required.

Therefore we are dealing with the inverse problem in this case. Thus in this part of the problem, we are looking for a value of the Z variable for which the entire area to the left is equal to 0.9834. If we use the normal chart, then we look in the body of the chart and locate the given value, i.e., 0.9834 or if the chart that we are using provides only the area between 0 and a value of Z, then we locate 0.4834, which is the given area subtracted by 0.5. Then, we read off the corresponding value of Z in the first column of the chart. Here, we see that the answer is

$$c = 2.13$$

Once again, modern calculators provide a function for solving the inverse problem. In the Texas Instruments calculators, this function is *invNorm*. The function requires one argument and that is the total area to the left of the required Z value. Thus in this case, we could solve the problem by using

$$c = invNorm(0.9834) = 2.13$$

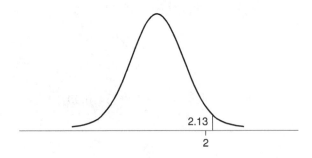

SOLUTION (Part k):

As discussed before, the 90th percentile of a distribution corresponds to a value of the variable for which 90% of all values of the variable are below it. Therefore stated in another way, the 90th percentile of the standard normal variable is a value c such that

$$P(Z < c) = 0.90$$

Thus by using either the normal distribution table or the calculator, we have

$$c = invNorm(0.90) = 1.28$$

Exercise 2

SHOW ALL WORK. For the standard normal random variable Z compute

a. $P(0 < Z < 2.03)$

b. $P(Z > -1.77)$

c. $P(1.22 < Z < 2.21)$

d. A constant c such that $P(Z > c) = 0.8749$.

e. The third quartile of the distribution

Section 4: General Case—Nonstandard Normal

As explained before, the importance of the normal distribution stems from its application. There are many random phenomena in nature that can be modeled as a normal random variable. However, in practice, we probably never encounter a situation where the underlying variable has a standard normal distribution. So, do we find the probabilities and consequently the area under the normal curve for a nonstandard normal distribution when the mean is not 0 or the standard deviation is not 1? The answer is actually not difficult. The reason that we discussed the standard normal distribution first is that there is a very close tie between any nonstandard normal variable and the standard normal variable. This close tie is defined by the Z score that we also met in Chapter 2. In fact if X is a normal random variable with mean μ and standard deviation σ, then its Z score

$$z = \frac{X - \mu}{\sigma}$$

has a standard normal distribution. Using this relationship, we can convert a nonstandard normal problem to a standard one and solve as before. So, for example, if $\mu = 100$ and $\sigma = 10$ say, then

$$P(83 < X < 122) = P\left(\frac{83 - 100}{10} < Z < \frac{122 - 100}{10}\right) = P(-1.7 < Z < 2.2)$$

and as before, we can use either a normal distribution table or a calculator to compute this probability. In particular in this case we have

$$P(-1.7 < Z < 2.2) = normalcdf(-1.7, \ 2.2) = 0.9415.$$

Modern calculators, however, have made it even easier for us. The *normalcdf* function on the TI calculators can also provide the area between two values for a nonstandard normal distribution as well. Although the default is the standard normal, to calculate the area under the normal curve for a nonstandard distribution, we have to supply the values of the mean and standard deviation. In other words to calculate the area under a normal distribution with mean μ and standard deviation σ between two numbers a and b, we enter *normalcdf* (a, b, μ, σ) If the last two values are ignored, the calculator assumes that μ is 0 and σ is 1 as in the standard normal. Therefore in the above example, we could find the probability using

$$P(83 < X < 122) = normalcdf(83,122,100,10) = 0.9415$$

Similarly, in solving the inverse problems, we can find the Z score and use the normal distribution chart. For example, suppose for the normal distribution with $\mu = 100$ and $\sigma = 10$, we are interested to find the 90th percentile of the distribution, i.e., we are seeking a constant c such that

$$P(X < c) = 0.90$$

Then, converting to the standard normal variable, we have

$$P\left(Z < \frac{c - 100}{10}\right) = 0.90$$

and using a normal distribution table, we see that although 0.90 is not exactly in the body of the chart, the closest value 0.8977 corresponds to a Z value of 1.28. Alternatively, using the *invNorm*(0.90) = 1.28 on our calculator, we have

$$(c - 100)/10 = 1.28$$

Upon solving the above equation for c, we have $c - 100 = 12.8$, from which we obtain $c = 112.8$. However, once again, our calculator makes this process easier for us. The *invNorm* function can solve the inverse problem for the nonstandard normal distribution once we provide the mean and standard deviation of the distribution. The default is the standard normal, but if we want to solve the inverse problem and find a value of a nonstandard normal random variable X for which the total area to the left under the normal curve is say p, then we can enter *invNorm*(p, μ, σ). Thus in the above example, to find the 90th percentile of a normal random variable for which for which $\mu = 100$ and $\sigma = 10$, we find that

$c = $ *invNorm*$(0.90, 100, 10) = 112.8$.

■ Example 5: Probability Calculation for the General (Nonstandard) Normal Distribution

The lifetime of a brand of light bulbs is known to have a normal distribution with a mean of 253 hours and standard deviation of 74 hours. If you purchase one light bulb of this brand, what is the probability that it would last

a.

    i. Less than 150 hours

    ii. More than 400 hours

    iii. Between 100 and 180 hours

b. What proportion of light bulbs last longer than 450 hours?

c. What lifetime value is exceeded by 80% of the light bulbs?

d. What is the 90th percentile of the lifetime distribution?

e. If the company wants to set a warranty limit, what lifetime value should be considered so that no more than 5% of product is returned for a refund?

SOLUTION: Here, the mean μ is 253 and the standard deviation σ is 74. Therefore,

a.

  (i) $P(X < 150) = $ *normalcdf* $(-\infty, 150, 253, 74) = 0.0820$

  (ii) $P(X > 400) = $ *normalcdf* $(400, +\infty, 253, 74) = 0.0235$

  (iii) $P(100 < X < 180) = $ *normalcdf* $(100, 180, 253, 74) = 0.1426$

b. $P(X > 450) = $ *normalcdf* $(450, +\infty, 253, 74) = 0.0039$

c. This part is essentially asking for the 20th percentile. Thus

  $c = $ *invNorm* $(0.20, 253, 74) = 190.72$ *hours*.

d. Similarly, for the 90th percentile, we have

  $c = $ *invNorm* $(0.9, 253, 74) = 347.83$

e. Here, we are looking for a lifetime value such that 5% of light bulbs fail to reach that value, i.e., the 5th percentile. Thus, we have

  $c = $ *invNorm* $(0.05, 253, 74) = 131.28$ *hours*.

Exercise 3

A survey was conducted in a city to study the commuting time to work. It was found that the average commuting time was 43.2 minutes with a standard deviation of 12.8 minutes. Assuming that the commuting time X is a variable that has a normal distribution,

a. What proportion of commuters take longer than an hour to commute to work?

$P(X > 60) =$

b. What is the probability that a commuter would have less than 15 minutes commute to work?

$P(X < 15) =$

c. If the commuting population of the city is about 45,000, approximately how many people take between 20 minutes and 70 minutes to get to work?

$P(20 < X < 70) =$

Approximate Number =

d. 90% of commuters take less than how long to get to work?

$c =$

e. What is the lower quartile of the commuting time distribution?

■ Example 6: Grading on a curve

Suppose that grades in an exam have a normal distribution with mean 72 and standard deviation of 8. Determine cut-off values for letter grades so that the top 10% get As, the next 15% get Bs, the next 40% get Cs, the next 30% get Ds, and the remaining 5 % get Fs.

For the top 10%, we need a value such that 90% of scores exceed that value, i.e., the 90th percentile. Thus the grade A cut-off is determined as

$$c = invNorm(0.90, 72, 8) = 82.25$$

For the next 15%, we need a value so that 25% of scores exceed that value, i.e., the 75th percentile. Thus for grade B we have

$$c = invNorm(0.75, 72, 8) = 77.39$$

Similarly, for the next 40% we need a value exceeded by 65% of scores, i.e., the 35th percentile, that is, the cut-off for grade C is

$$c = invNorm(0.35, 72, 8) = 68.92$$

Finally for grade D we need the 5th percentile,

$$c = invNorm(0.05, 72, 8) = 58.84.$$

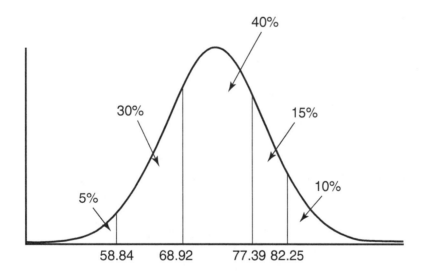

Exercise 4

The distribution of the typing rate measured by the number of words per minute for experienced typists is approximately normal with a mean of 64 words per minute with a standard deviation of 14 words per minute.

a. How many words per minute do you expect the fastest 10% type?

$c =$

b. What are the lower and upper quartiles of the distribution?

$Q_1 =$

$Q_3 =$

c. How many words per minute do the slowest 5% type?

$c =$

Exercise 5

The average height of college students is about 69.2 inches with a standard deviation of 2.8 inches. Assuming that the height of all college students has a normal distribution. Among 5000 students, how many would you expect to have heights that

a. Exceeds 6′ 2″?

b. Is between 6′ and 7″?

c. If the height is measured to the nearest 0.5 inch, how many would you expect to have heights equal to 5′ 7″?

d. Determine the quartiles and the interquartile range.

Exercise 6

A healthy range for the blood glucose level is considered to be between 70 and 99 mg/dL. Assuming that the blood glucose levels have a normal distribution with a mean of 85 mg/dL and a standard deviation of 8.3 mg/dL,

a. What proportion of population is considered to have normal blood glucose level?

b. If blood glucose levels above 120 are considered to be abnormally high, what proportion of the population has abnormally high blood glucose level?

c. We wish to identify the 10% of the population with highest blood glucose level for some educational training. What is the cut-off value for this group of individuals?

Chapter 6

Sampling Distribution

Section 1: Introduction

In previous chapters, we learned that one of the duties of a statistician is to utilize the information in the sample in order to learn about the population characteristics. Often, when we are interested to learn about say, the average of a variable, we take a sample and compute the mean of the sample in order to use it as an estimate for the population. For example, if you wish to know how many hours on average students in your university spend per week studying outside the classroom, you take a random sample of say 50 students at your university and from each selected student you ask the same question about the number of hours they spend studying outside the classroom. You then take the 50 numbers and average them, and use the average as your estimate for the average number of hours all students in your college spend studying outside the classroom. Similarly, if a public health official in our region wants to estimate the proportion of people in the region who get flu shots every winter, he/she will take a random sample and based on the sample proportion an estimate for the entire population is derived. This methodology is, of course, fine except that we should remember that the estimates are based on a single sample. If the process of sampling is repeated, the estimates are likely to be different. In fact each time the process of sampling is repeated, the value of a statistic computed from the data may change. This is due to what is called *Sampling Variability*. In this chapter, we study this sampling variability and try to see how we can characterize the changes that may occur in the value of a statistic calculated from a sample in repeated sampling. The characterization of fluctuation of a statistic in repeated sampling is called the *Sampling Distribution* of that statistic. Although we mainly concentrate on the sampling distribution of the sample mean, we should know that a sampling distribution exists for any statistic that is computed from the sample. In order to clarify the idea of sampling variability and sampling distribution, we begin with a simple example and go on to consider more general cases. We will discuss a very important theorem in statistics that gives us the sampling distribution of the sample mean under very general conditions.

■ Example 1

Suppose that a bag contains 4 balls numbered 1–4. Two balls are drawn at random and with replacement.

- **a.** List all possible samples.
- **b.** Compute the mean for each sample.
- **c.** Obtain the sampling distribution for the sample mean.
- **d.** Compute the mean and standard deviation for both the parent population from which the sample is taken and also for the sampling distribution of the mean and compare.

Here, because the bag has only 4 balls and each number has the same chance as other numbers, the probability distribution of the population from which the sample is taken, is as follows:

x	1	2	3	4
$P(x)$	$\dfrac{1}{4}$	$\dfrac{1}{4}$	$\dfrac{1}{4}$	$\dfrac{1}{4}$

Therefore we have a population of size 4, from which we wish to take a sample of size 2. Below is a list of all possible samples, where for each sample we have also calculated the sample mean by simply adding the two numbers and dividing the result by 2.

Sample	Mean	Sampl	Mean
(1,1)	1	(3,1)	2
(1,2)	1.5	(3,2)	2.5
(1,3)	2	(3,3)	3
(1,4)	2.5	(3,4)	3.5
(2,1)	1.5	(4,1)	2.5
(2,2)	2	(4,2)	3
(2,3)	2.5	(4,3)	3.5
(2,4)	3	(4,4)	4

Note that each sample consists of a pair, the first number indicating the number on the ball drawn out of the bag first, and the second number showing the ball that is drawn the second time. The above table shows that there are 16 possible samples and depending on which sample is drawn from the bag, the value of the sample mean can be different. This is due to sampling variability. Some values of the mean such as 2.5 have a higher chance of occurring whereas other values such as 1.0 and 4.0 can occur only when the ball with number 1 or the ball with number 4 is drawn twice. Now, because the sampling is completely random, all samples have the same chance of occurrence and indeed each sample has 1/16 chance of being drawn. Hence, if we collect the information from the above table, we get the following distribution, which is called the sampling distribution of $\bar{X}$:

Sampling distribution of $\bar{X}$

$\bar{X}$	1.0	1.5	2.0	2.5	3.0	3.5	4.0
$P(\bar{X})$	1/16	2/16	3/16	4/16	3/16	2/16	1/16

Now, based on what we learned in Chapter 4, let us compute the mean and standard deviation for both of the above distributions. First, for the population distribution we have

X	1	2	3	4	
$P(X)$	1/4	1/4	1/4	1/4	$\mu = 2.5$
$X.\,P(X)$	1/4	1/2	3/4	1	
$(X-2.5)^2 * P(X)$	9/16	1/16	1/16	9/16	$\sigma^2 = 1.25$

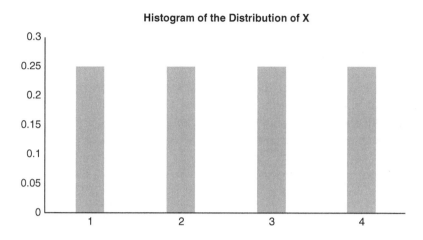

Histogram of the Distribution of X

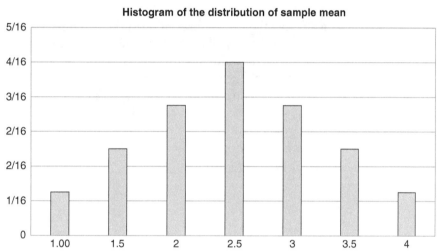

Histogram of the distribution of sample mean

and similarly for the sampling distribution, we have

X	1	1.5	2	2.5	3	3.5	4
$P(X)$	1/16	1/8	3/16	1/4	3/16	1/8	1/16
$X.P(X)$	1/16	3/16	3/8	5/8	9/16	7/16	1/4
$(X-2.5)^2 P(X)$	9/64	1/8	3/64	0	3/64	1/8	9/64

$$\mu_{\bar{X}} = 2.5, \ \sigma_{\bar{X}}^2 = 0.625$$

Now, there are a few interesting facts that we should note about the above distributions. First, we see that the mean of the $\bar{X}$ sampling distribution is equal to the mean of the distribution from which the sample was drawn. Is that a coincidence? Well no, it turns out that we can prove mathematically that when samples of the same size are drawn repeatedly from a distribution, the means of the samples will have a distribution that is centered on the same point as the parent population, i.e., we have

$$\mu_{\bar{X}} = \mu$$

for any distribution. This is a very interesting fact that makes the sample mean an attractive candidate for estimating the population mean. We will talk about estimation in the next chapter, but here we note that what this means is that when we use the sample mean to try and estimate the population mean, even though we may not get the exact value of the population mean, but if the process of sampling is repeated enough number of times, then on average we hit the right value for estimating the population mean. This property of the sample mean is called unbiasedness and we call the sample mean an *unbiased* estimate of the population mean.

The second thing that we note from the above two distributions is the relation between the variances. Although the two variances are not the same, they are closely related and the variance of the $\bar{X}$ sampling distribution is exactly equal to the variance of the parent distribution divided by the number 2. But, why 2? Where does the number 2 come from and how does it relate to this problem? Well, we notice that the size of the samples drawn from the parent population was 2 and indeed this is another general result that can be proven mathematically in more advanced statistical courses. In general, we can prove that when samples of size n are taken from a population with variance σ^2, then the variance of the $\bar{X}$ sampling distribution is precisely σ^2/n. This, of course, in turn means that we can derive a formula for the standard deviation of the $\bar{X}$ sampling distribution by simply taking the square root of this expression. Thus we have

$$\sigma_{\bar{X}} = \frac{\sigma}{\sqrt{n}}$$

Note that the above formula indicates that the standard deviation of the $\bar{X}$ sampling distribution decreases as we increase the sample size. In other words, it says that the sampling distribution of the sample mean is more concentrated (less variation) around the true value μ in larger samples. This, of course, makes perfect sense that in larger samples, we are more likely to have more accurate estimates for the population mean. In fact the above equation shows that if we increase the sample size by any factor, the standard deviation of the $\bar{X}$ sampling distribution is reduced by the square root of that factor. For example, if we increase the sample size by a factor of 4, then the standard deviation of the $\bar{X}$ sampling distribution is reduced by a factor of 2. Similarly, if we increase the sample size by a factor of 9, the standard deviation of the sampling distribution of the sample mean is reduced by a factor of 3 and so on.

Exercise 1

A random sample of 10 items is taken from a population that has a mean of $\mu = 12.5$ and a standard deviation of $\sigma = 8.3$. Compute the mean and standard deviation of the sampling distribution of the sample mean. If we desire to reduce the standard deviation of the sample mean by a factor of 2, what sample size should be used in the study?

Through the example in this section, we were able to explore the notion of sampling distribution and also discover some of the interesting properties of the sampling distribution of the sample mean. The example is a very simple one using samples of size 2 drawn from a population of size four. In practice, we usually deal with populations that have larger sizes and our samples are normally higher than size 2. In the next section, we discuss the sampling distribution of the sample mean in more general cases.

Section 2: Sampling Distribution of the Sample Mean $\bar{X}$

Suppose that a random sample of size n is taken from a population with mean μ and standard deviation σ. Let $\bar{X}$ be the sample mean. Clearly in repeated sampling, the value of $\bar{X}$ varies from one sample to another. In fact each time the process of sampling is repeated, the value of the sample mean may be different. Characterization of this fluctuation gives the sampling distribution of $\bar{X}$. How can we characterize the sampling distribution of $\bar{X}$? In order to answer this question, we consider two cases:

First, for simplicity, we only consider the normal distribution case, i.e., we consider the situation in which the distribution of the parent population is normal and samples of same size n are being drawn from this normal distribution. Let's look at an example:

Example 2

The monthly premium paid for auto insurance in a certain region has a normal distribution with a mean of $120 and a standard deviation of $15. We want to do a sampling study with this distribution. Suppose we take a sample of 5 individuals and from each individual find out how much they pay for their auto insurance per

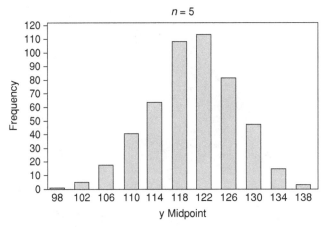

Figure 6.1: $n = 5$

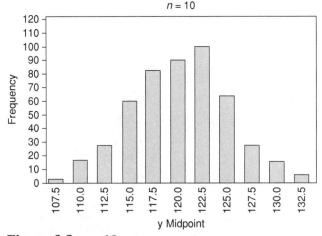

Figure 6.2: $n = 10$

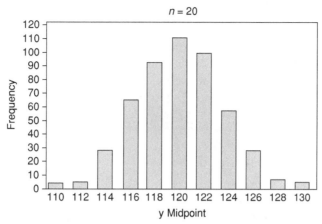

Figure 6.3: $n = 20$

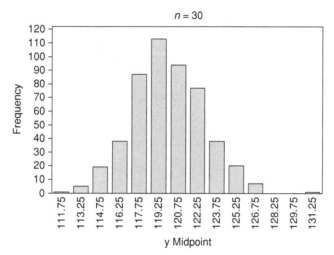

Figure 6.4: $n = 30$

month. We find the mean of the sample and repeat the process 500 times. We then sketch a frequency distribution chart to examine the sampling distribution of the mean. In practice, this process is easily done with computer simulation. Figure 6.1 shows the sampling distribution of the sample mean for samples of size 5. Figures 6.2, 6.3, and 6.4 display the same information, i.e., the sampling distribution of the sample mean, but for samples of size 10, 20, and 30, respectively. The SAS computer software was used to make the simulation and generate the charts. Interestingly, we see that every time, no matter the sample size, the shape of the sampling distribution of the sample mean resembles a normal curve and all the distributions are centered around 120 which is the mean of the population from which the samples were taken. The difference is however, that as we increase the sample size, the associated normal curve seems to become narrower. This in turn means that the standard deviation of the sampling distribution decreases. Thus we have the following theorem:

❖ **Theorem 6.1:** If $\bar{X}$ is the mean of a sample of size n from a population whose distribution can be characterized by a normal curve with mean μ and standard deviation σ, then the sampling distribution $\bar{X}$ has the following properties for any value of n (all sample sizes):

- The sampling distribution is centered at population mean $\mu_{\bar{X}} = \mu$
- Standard deviation of the sampling distribution decreases as sample size increases

$$\sigma_{\bar{X}} = \frac{\sigma}{\sqrt{n}}$$

- Shape of the distribution is normal for any sample size.

■ Example 3

We consider once again the lifetime of light bulbs example that we saw in Chapter 5. In that example we said that it was known that the lifetime of a brand of electrical light bulbs has a normal distribution with a mean of 253 hours and a standard deviation of 74 hours. Suppose now that you purchase 12 of these light bulbs. What is the probability that the average lifetime of these 12 light bulbs:

    a. Exceeds 300 hours?
    b. Is between 200 and 220?
    c. What is 90th percentile of the $\bar{X}$ distribution?

SET UP: Here, we are sampling $n = 12$ bulbs from a normal distribution for which $\mu = 253$ and $\sigma = 74$. Thus from the above theorem (Theorem 6.1) we can say that $\bar{X}$ will have a normal distribution with

$$\mu_{\bar{X}} = 253 \text{ and } \sigma_{\bar{X}} = \frac{74}{\sqrt{12}} = 21.36$$

SOLUTION:

Hence, we have

    a) $P(\bar{X} > 300)$

$$= P\left(Z > \frac{300 - 253}{21.36}\right)$$

$$= P(Z > 2.20)$$

$$= 0.5 - 0.4861 = 0.0139$$

Or simply $0.5 - \text{normalcdf}(253,300,253,21.36)$

    b) $P(200 < \bar{X} < 220) = P\left(\frac{200 - 253}{21.36} < Z < \frac{200 - 253}{21.36}\right)$

$$= P(-2.48 < Z < -1.54) = 0.4934 - 0.4382 = 0.0552$$

Similarly $\text{normalcdf}(200,220,253,21.36)$

    c) $P(\bar{X} < c) = 0.90$ and thus $P\left(Z < \frac{c - 253}{21.36}\right) = 0.90$

$$\frac{c - 265}{21.36} = 1.28 \text{ hours}$$

$$c = 280.34 \text{ hours}$$

which could also be derived using InvNorm(0.9,253,21.36).

We now return to the question of sampling distribution of the sample mean in the more general case when the sample is taken from a population whose distribution is not symmetric and does not resemble the normal curve. Once again we consider an example.

Exercise 2

In Exercise 3 of Chapter 5, we considered a survey regarding commuting time to work for a certain city. The results of the survey had indicated that the distribution of commuting time was normal with a mean of 43.2 minutes and standard deviation of 12.8 minutes. For a group of 20 commuters in this city:

a. What is the probability that their average commute time is 40 and 50 minutes? Longer than one hour?
b. What is the 90th percentile of the distribution of the average commuting time for all groups of 20 people?

■ Example 4

In a certain community, the distribution of the daily consumption of water in millions of liters has a mean 1 and a standard deviation of 5. The relatively large standard deviation of this distribution clearly indicates that the distribution is not symmetric. In fact if we subtract twice the standard deviation from the mean, we get a negative number which, of course, is meaningless for water consumption. For normal distributions, on the other hand, two standard deviations below the mean are quite meaningful and as we know, according to the empirical rule about 95% of measurements are within two standard deviations and therefore quite feasibly, about 2.5% of measurements are expected to fall below two standard deviations. In fact the graph of this distribution is given below, where we can see how the distribution is positively skewed.

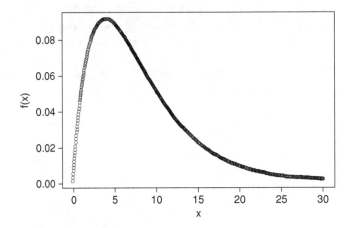

Once again, we perform a simulation study on this distribution to examine the shape of the sampling distribution of the sample mean for different sample sizes.

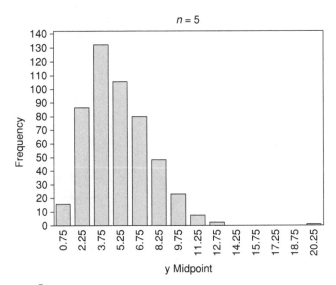

$n = 5$

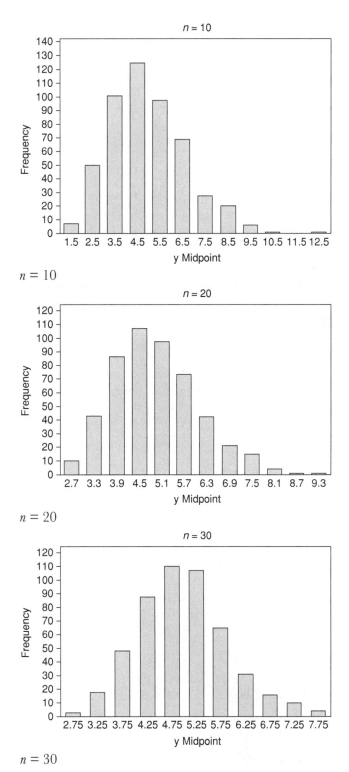

$n = 10$

$n = 20$

$n = 30$

The above figures show the sampling distribution of the sample mean when samples of size 5, 10, 20, and 30 are taken from the parent distribution. So notice that as the sample size increases, the shape of the sampling distribution gets closer and closer to the normal curve. Remarkably, we see that even though the shape of the parent distribution is completely non-normal, but when the sample size is 30, the shape of the sampling distribution of the sample mean very much resembles a normal distribution. This interesting result has very important applications and because of its importance, it has a name and it is known as the Central Limit Theorem. In the eyes of some statisticians, the Central Limit Theorem is the most important result is statistics.

❖ Theorem 6.2: The Central Limit Theorem

If $\bar{X}$ is the mean of a sample of size n from a population with mean μ and standard deviation σ, then the sampling distribution has $\bar{X}$ the following properties:

- The sampling distribution is centered at population mean $\mu_{\bar{x}} = \mu$
- Standard deviation of the sampling distribution decreases as sample size increases

$$\sigma_{\bar{x}} = \frac{\sigma}{\sqrt{n}}$$

- Shape of the distribution approaches that of the normal distribution as sample size increases.

As mentioned above, the importance of the Central Limit Theorem is in its application. The ramification of the Central Limit Theorem is that if the sample size is sufficiently large, we approximate the shape of the sampling distribution of sampling distribution $\bar{X}$ with normal. In practice, it has been shown that in most situations, even a sample size of size 30 is sufficient to have a sampling distribution that is fairly close to a normal, although in some cases even smaller sample sizes may suffice. Thus as a rule of thumb, whenever the sample size is 30 or larger, i.e., $n \geq 30$, we treat it as a large sample and apply the Central Limit Theorem.

■ Example 5: Bus delay

We return to an example that we considered earlier in the last chapter for the uniform distribution. We said that a bus is scheduled to arrive at a station at every hour on the hour. However, buses in the evening can be late between 0 and 10 minutes according to a uniform distribution. Over a period of one month, what is the probability that average delay

a. Exceeds 6 minutes?
b. Is less than 3 minutes?
c. What is 90th percentile of distribution of average delay?

SET UP: Now, in this problem, we are not interested in the delay of the bus on a single day. Rather, we want to know about the average delay over 30 days. Thus we can treat the delays as a sample of size 30 from the uniform distribution. Although the parent distribution is uniform and looking completely non-normal, since we have a sample that is large enough, we can apply the Central Limit Theorem and say that the sampling distribution of $\bar{X}$ is approximately normal with mean and standard deviation respectively given by

$$\mu_{\bar{x}} = 5 \qquad \sigma_{\bar{x}} = \frac{2.89}{\sqrt{30}} = 0.53$$

SOLUTION:

Therefore, we have

a. $P(\bar{x} > 6) = P\left(z > \frac{6-5}{0.53}\right) = P(z > 1.89) = 0.5 - 0.4706 = 0.0294$

or $0.5 - normalcdf(6, \infty, 5, .53) = 0.0294$.

b. $P(\bar{x} < 3) = P\left(z < \frac{3-5}{0.53}\right) = P(z < -3.77) = 0$

Or $normalcdf(-\infty, 3, 5, 0.53)$.

c. $P(\bar{x} > c) = .90 \quad P\left(z < \frac{3-5}{.53}\right) = 0.9 \quad \frac{c-5}{0.53} = 1.28 \quad c = 5.6784$ minutes

Which could also be derived using $invNorm(0.9, 5, 0.53) = 5.6784$.

Exercise 3

It is known that the distribution of the blood sugar level of individuals under 50 in a certain geographic area has a mean 85 and a standard deviation of 25. Let $\bar{X}$ be the mean of the blood sugar level for a random sample of 45 individuals. Find the probability that for the sample, the average blood sugar level is:

a. Below 80
b. In excess of 142
c. Between 112 and 150

In 90% of samples of size 45, the blood sugar level is below what value?

Statistical Inference

Section 1: Introduction

Statistical inference is the branch of statistics that utilizes the information in the sample to make meaningful statements about unknown characteristics of a population. These unknown characteristics are often stated in terms of some parameters such as the population mean, the population variance, or some other parameters. This branch of statistics is also concerned about evaluating the reliability of the derived statements. There are generally two types of inferential problems in statistics. One is estimation where we try to determine plausible values for unknown population parameters based on samples. The other type of inferential problem is hypotheses testing. In these problems, the sample is used to verify the validity of a claim. For example, if a certain pharmaceutical company claims that their new drug has a superior cure rate compared with all similar drugs, how can we go about testing the validity of this claim. Similarly, if a manufacturer claims that their brand of product has an average lifetime that is better than all other brands, how do we examine the validity of this statement. In this chapter, however, we study the estimation problems and in the next chapter we discuss hypotheses testing problems.

When it comes to estimation of unknown parameters in a population, there are two types of estimates that one can envisage. One is called point estimation when a single value of the parameter is determined and the other is interval estimation, where instead of a single value, a whole range of possible values of the parameter is derived. The advantage of this type of estimation is that a level of confidence is also attached to the interval. For example, we might state that we estimate the average life of a product is between two numbers a and b with say, 95% confidence. That is why this latter type of estimation is often called 'confidence intervals.' In this chapter, we first discuss point estimation and then introduce the confidence intervals.

Section 2: Point Estimate

A point estimate of an unknown population quantity is a single number derived from the sample which gives a plausible value for the population quantity. We have seen, for example, that when we want to learn about the average of a certain quantitative characteristic for an entire population, very often the average of that characteristic in a random sample provides a good estimate, provided that the sample size is moderately large. The fact is that sample quantities have many desirable properties in estimating unknown population values. Thus in order to give a point estimate of an unknown population quantity, we can simply use the equivalent sample quantity. The table below gives the point estimate of some of the population parameters that we have seen in the past:

Population Quantity	Point Estimate
Mean μ	Sample mean $\bar{X}$
Median	Sample median
Variance σ^2	Sample variance S^2
Standard deviation σ	Sample standard deviation S
Range	Sample range
Quartiles	Sample quartiles
. . .	. . .

■ Example 1: Application in Experimental Psychology

To determine the behavioral effect of a chemical, in a psychology experiment, eight rats were injected with a dose of the chemical and time to a certain reaction (e.g., returning to cage) was determined for each rat in minutes,

$$1.8 \quad 2.5 \quad 2.3 \quad 3.1 \quad 2.8 \quad 3.2 \quad 1.7 \quad 2.1$$

a. Estimate mean reaction time to the drug for all rats exposed to the chemical.
b. Give a point estimate of standard deviation of reaction time.
c. Assuming a normal distribution for reaction times, give a point estimate for the 90th percentile of the distribution.
d. Give an estimate for probability that reaction time is longer than 3 minutes.

SOLUTION
a. We estimate the population reaction time using the sample mean. Thus

$$\bar{X} = 2.4375 \text{ minutes}$$

b. Similarly to estimate the population standard deviation, we calculate the sample standard deviation. Accordingly, we first compute the sample variance. We have

$$\sum X = 19.5 \quad \sum X^2 = 49.77$$

$$S^2 = \frac{49.77 - \frac{(19.5)^2}{8}}{7} = 0.3198 \text{ min}^2$$

and therefore

$$S = 0.5655 \text{ minutes}$$

c. Using the normal distribution and the point estimates of the mean and standard deviation, we seek c such that

$$P(X < c) = .90$$

or

$$P\left(Z < \frac{c - 2.4375}{.5655}\right) = .90$$

and so since $invNorm(0.9) = 1.28$, we have

$$\frac{c - 2.4375}{0.5655} = 1.28$$

or

$$c = 3.1614 \text{ minutes}$$

d. Lastly

$$P(X > 3) = P\left(Z > \frac{3 - 2.4375}{.5655}\right) = P(Z > .99) = .5 - .3389 = .1611$$

YOUR TURN! Please complete the required exercises below directly in this book. You are encouraged to discuss each exercise with a partner or group. You may be asked to complete these exercises in class or outside of class.

Exercise 1

Table below gives the average annual rainfall for eight cities in Florida over the period of 1981 to 2010.

City	1981–2010
Orlando	50.73″
Daytona Beach	49.62″
Fort Myers	55.93″
Key West	39.83″
Miami	61.90″
Tampa	46.30″
Jacksonville	52.39″
Tallahassee	59.23″

a. Give a point estimate of the average annual rainfall for all cities in Florida

b. Give a point estimate for the standard deviation of the amount of annual rainfall

c. Give a point estimate for the median of the annual rainfall

d. Assuming a normal distribution, give a point estimate for the 90th percentile of the annual rainfall in Florida

e. Using a normal distribution, give an estimate of the probability that the annual rainfall in Florida exceeds 55 inches.

$$\sum X = 415.93$$

$$\sum X^2 = 21978.4761$$

Section 3: Interval Estimation

An interval estimate of an unknown population quantity is a range of plausible values of that quantity determined from a random sample with a certain level of confidence. Although in practice there may be any number of different parameters in a population that are unknown and need to be estimated, here we only consider interval estimation of the population mean μ. Also, in discussing the interval estimation of the population mean, we consider two cases. First, we consider a situation where it is possible to draw a large sample from the population and next we discuss situations where due to the experimental conditions and restrictions such as cost, time, etc., it is not possible to draw a random sample of large size from the population. The reason for this distinction is that, as we saw in Chapter 6, when it is possible to have a large sample from the population, then we have this powerful tool, namely the Central Limit Theorem that can provide a good approximation for the sampling distribution of the sample mean. When the sample size is small, the Central Limit Theorem of course does not apply.

Large Sample ($n \geq 30$) Confidence Interval for μ.
Suppose that a random sample of size n (large) is given from a population with mean μ and standard deviation σ. It is desired to obtain an interval estimate for μ. Let $\bar{X}$ and S respectively be the sample mean and standard deviation. Then, since n is large, we know that by the Central Limit Theorem, $\bar{X}$ will have approximately a normal distribution with

$$\mu_{\bar{X}} = \mu, \qquad \sigma_{\bar{X}} = \sigma / \sqrt{n}$$

Therefore we can say that

$$Z = \frac{\bar{X} - \mu}{\sigma / \sqrt{n}}$$

has a standard normal distribution. Suppose it is desired to derive a 95% confidence interval. Using the normal distribution table or the calculator, we can determine the two values that embrace 95% of the area in the middle of the normal curve. In fact using the invNorm function on our calculator, we get invNorm(0.025) = −1.96 and invNorm(0.975) = 1.96. That is

$$P(-1.96 < Z < 1.96) = 0.95$$

Replacing Z by the expression from above, we have

$$-1.96 < \frac{\bar{X} - \mu}{\sigma / \sqrt{n}} < 1.96$$

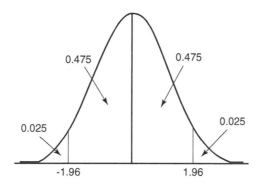

Solving for μ, we get the formula for the 95% confidence interval as

$$\overline{X} - 1.96\frac{\sigma}{\sqrt{n}} < \mu < \overline{X} + 1.96\frac{\sigma}{\sqrt{n}}$$

Note: The above equation contains the population standard deviation σ and in practice σ is often unknown. In such cases, we use its point estimate S and argue that since the **sample size is large**, the point estimate of σ is likely to be very close to the true population value and this substitution will not affect the confidence interval significantly.

■ Example 2: Pharmaceutical Application

In order to estimate the average time that a drug takes to be absorbed in the bloodstream, a pharmaceutical company uses a random sample of 40 individuals as test samples. After giving the drug to each patient, time to complete absorption is measured for each patient. It is found that for the sample, the average time to absorb the drug is 38 minutes with a standard deviation of 18 minutes. We wish to estimate the average time for absorption of the drug for all patients taking this drug. Using a 95% confidence interval, we have

$$n = 40 \quad \overline{X} = 38 \quad S = 18$$

$$38 - 1.96 \cdot \frac{18}{\sqrt{40}} < \mu < 38 + 1.96 \cdot \frac{18}{\sqrt{40}}$$

$$32.42 < \mu < 43.58 \text{ minutes.}$$

This means that we are 95% confident that the mean time to absorption of the drug is a number between 32.42 and 43.58.

Meaning of Confidence Interval: In the above example, exactly what does it mean to say that we are 95% confident? How do we interpret confidence in this case? Well, first note that both sides of the formula for the 95% confidence interval above depend on $\overline{X}$. Thus, because $\overline{X}$ changes from one sample to another, (it is a random variable) this means that both sides of the confidence interval will change each time the experiment is repeated with a new sample. The derived interval may or may not contain the true value of the population mean μ. Of course we don't know. However, when we construct a say 95% confidence interval, it means that if the process is repeated a large number of times, then in about 95% of the intervals we capture the true value of the population mean.

Exercise 2

A large company is interested to estimate the mean number of days that its employees miss due to sickness during a certain season. From the past records, a random sample of 36 is selected and the number of sick leaves for each record is determined. It is found that for the sample, the mean was 9.4 days with a standard deviation of 5.8 days. Estimate the mean number of sick leave days for all employees of this company using a 95% confidence interval. Interpret the interval.

Section 4: Varying Confidence Levels

Now, suppose that rather than a 95% confidence interval, in general, a confidence interval is desired for μ with a level of confidence $1 - \alpha$. Since n is large, as before,

$$Z = \frac{\overline{X} - \mu}{\sigma / \sqrt{n}}$$

is a standard normal variable and we seek two numbers a and b such that

$$P(a < Z < b) = 1 - \alpha$$

That is to say that the tail area in the normal curve is some small number, say, α. We split this value equally for the two tails so that the area for each tail is $\alpha/2$. Once again, using the normal distribution table or the invNorm function on our calculator, we can determine the values on the horizontal axis that correspond to these tail areas. By symmetry of the normal distribution, of course the two values are the same distance from the center. So, let's call them $-Z_{\alpha/2}$ and $Z_{\alpha/2}$.

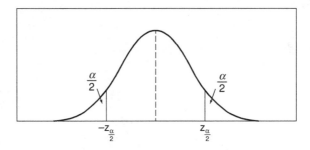

Thus $a = -Z_{\alpha/2}$ and $b = Z_{\alpha/2}$ and with a procedure similar to what we saw for the 95% confidence interval, we find the following formula for the general case of the confidence interval.

$$\overline{X} - Z_{\frac{\alpha}{2}} \cdot \frac{\sigma}{\sqrt{n}} < \mu < \overline{X} + Z_{\frac{\alpha}{2}} \cdot \frac{\sigma}{\sqrt{n}}$$

The quantity $Z_{\frac{\alpha}{2}} \cdot \frac{\sigma}{\sqrt{n}}$ is called the **margin of error,** since it measures how much, on average our estimate may be off from the true value. In fact, the width of the confidence interval only depends on this quantity.

■ Example 3

Using the information in the previous example, we wish to compute 90% and 99% confidence intervals for mean absorption time of the drug. We then make a comparison of all three intervals:

For the 90% confidence, the area in the middle of the normal distribution is 0.9. Subtracting this value from 1 and dividing it by 2, we find each tail of the distribution has an area of 0.05. The values of the standard normal variable for which the tail area are 0.05, are −1.645 and +1.645. Thus, the 90% confidence interval is calculated as

$$38 - 1.645\frac{18}{\sqrt{40}} < \mu < 38 + 1.645\frac{18}{\sqrt{40}}$$

or

$$33.32 < \mu < 42.68 \text{ minutes.}$$

Similarly, for the 99% confidence interval, the area for each tail is 0.005 and the corresponding z-score is about 2.575. Thus we get

$$38 - 2.575\frac{18}{\sqrt{40}} < \mu < 38 + 2.575\frac{18}{\sqrt{40}}$$

or

$$30.67 < \mu < 45.55 \text{ minutes.}$$

as the desired confidence interval.

The Role of Confidence Level and Sample Size

Now, take a look at all three confidence intervals derived in examples 2 and 3. See what happens to the width of the confidence interval as we increase the level of confidence from 90% to 95% and to 99%. We can notice right away that as we increase the level of confidence, the width of the interval also increases. This means that as we demand a higher level of confidence from our estimate, we lose precision and have to accept more uncertainty. Does this make sense? Well, let's think about it. Our information is derived from the sample and since the size of the sample is fixed, the margin of error and consequently the width of the confidence interval depends solely on the level of confidence. The higher the level of confidence, the larger the margin of error. We have to balance the required level of confidence and the level of accuracy expected from the confidence interval. If we wish to have a higher level of confidence, we have to accept a wider interval, i.e., a lower level of accuracy unless we can add more information by taking more samples. In other words, we can design our experiment and select a sample size that according to our expectations and specifications will have a fixed level of confidence with a pre-specified width for the confidence interval. Note also that since the margin of error is inversely proportional to the square root of the sample size, to increase the accuracy and decrease the width of the confidence interval by a factor, the sample must be increased by the square of that factor. For example, if we wish to reduce the width of the confidence interval by one half, we need to quadruple the sample size and so on. We will discuss this issue of sample size determination a little later in this chapter. But, for now, we just mention that, as a rule of thumb, in most investigations, the 95% confidence interval is generally accepted as a reasonable trade off although that can change depending on the situation.

> **YOUR TURN!** Please complete the required exercises below directly in this book. You are encouraged to discuss each exercise with a partner or group. You may be asked to complete these exercises in class or outside of class.

Exercise 3

In the last exercise, you computed a 95% confidence interval for the mean number of sick day leaves of the employees of a company. Now, compute 80% and 98% confidence intervals and compare all three intervals.

Small Sample (*n* < 30) Confidence Interval for *μ*

There are many instances in practice that large samples are not available. A large sample is a luxury that is not afforded in all experimental situations. For example, in medicine, there are many times when physicians deal with rare diseases and need to collect information about the onset of the disease. Clearly, large samples are not feasible. In astronomy, sometimes scientists have to wait several years in order to make one observation. In engineering and life testing, often items have to be destroyed at the end of the experiment and large samples are not economical. Yet, as another example, in biological and animal bioassay experiments, often animals are exposed to toxic chemicals or are sacrificed at the end of the experiment and once again large samples are immoral. So, the question that we try to answer in this section is how we can construct confidence intervals for the mean when the sample size is small.

Suppose that a random sample of size *n* (small) is given from a population with mean *μ* and standard deviation *σ*. It is desired to obtain a confidence interval for *μ*. Let $\bar{X}$ and *S* be the sample mean and standard deviation, respectively. If the sample size is not large enough, clearly, the Central Limit Theorem does not apply. However, we learned in Chapter 6 that when the sample is drawn from a population with a normal distribution, then the sampling distribution of the sample mean also has a normal distribution no matter what the sample size. In other words, if the parent population follows a normal distribution, the sampling distribution of $\bar{X}$ also has a normal distribution for any sample size. Therefore, we are only going to concentrate on situations where the parent population from which the sample is taken has a normal distribution. Thus we make the following assumption:

Assumption: Sample is taken from a population whose density can be described (at least approximately) by a normal curve.

Then, sampling distribution of $\bar{X}$ is normal with mean and standard deviation respectively given by:

$$\mu_{\bar{X}} = \mu \qquad \sigma_{\bar{X}} = \sigma / \sqrt{n}$$

so that

$$Z = \frac{\bar{X} - \mu}{\sigma / \sqrt{n}}$$

is a standard normal variable.

One might think that at this point, we can apply the properties of the normal distribution similar to the large sample case and compute a confidence interval for the population mean *μ*. However, the statistic *Z* involves *σ* and in practice, the population standard deviation *σ* is generally an unknown parameter and we need to replace it with its point estimate. Earlier in this chapter, when we were constructing confidence intervals for *μ* in large samples, we encountered a similar situation and argued that since the sample size is large, replacing *σ* by its point estimate, that is the sample standard deviation *S*, will not affect the distribution of *Z*. Of course that argument cannot be made here since the sample size is small. In fact replacing *σ* by its estimate *S* will affect the distribution, and normality is no longer preserved. The statistic

$$T = \frac{\bar{X} - \mu}{S / \sqrt{n}}$$

does not have a normal distribution. This variable has the so called (Student) t-distribution with $(n-1)$ degrees of freedom. Since we have not seen this distribution before, below we provide a brief description of the t-distribution and some of its properties.

The t-distribution

The t-distribution is a symmetric, unimodal bell-shaped distribution. Compared to the normal distribution, the t-distribution has heavier tails than the normal curve. The t-distribution is characterized by one quantity called number of degrees of freedom. The higher the number of degrees of freedom, the closer the t-curve is to the normal curve. Below the graphs of the t-distribution for varying number of degrees of freedom is reproduced from the Internet where the parameter ν represents the number of degrees of freedom.

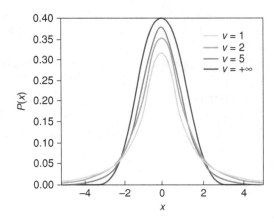

This distribution was first introduced by a British statistician by the name of w. S. Gosset who worked at the Guinness brewery. He used to publish his work under the name of Student. For this reason, the t-distribution is often called the Student's t-distribution.

Courtesy of Mehdi Razzaghi. Copyright © Kendall Hunt Publishing Company

Small Sample Confidence Interval for μ—Continued

Now, returning to the problem of finding the confidence interval for μ in small sample sizes, we see that in this case, rather than the normal distribution, the critical values of a t-distribution are used to construct the associated confidence interval. Specifically, if we desire a confidence interval with confidence coefficient of $1-\alpha$, we have

$$\bar{x} - t_{\alpha/2}\frac{S}{\sqrt{n}} < \mu < \bar{x} + t_{\alpha/2}\frac{S}{\sqrt{n}}$$

Where $t_{\alpha/2}$ represents the t critical value with $n-1$ degrees of freedom for which the tail area is $\alpha/2$. This value can be derived either from standard t-distribution tables or from scientific calculators. In the TI calculators, the function invT provides this critical value. For example, if a 95% confidence interval is desired and the sample size is say 15, then each tail area is 0.025 and the number of degrees of freedom is 14. Thus invT(.025, 14) and invT(.975, 14) would give us the two values of -2.145 and 2.145.

■ Example 4

A consumer protection agency is suspecting that the mean weight of the 60-pound potato bags packaged by a company is actually below 60. In order to estimate the mean weight of the packages, a random sample of 20 bags is selected and the weight of each bag is determined. It is found that average weight is 58 lb with a standard deviation of 8.3 lb. We wish to estimate the mean weight of all potato bags packaged by this company using

a. 90%
b. 98%

confidence intervals. Based on the intervals, is the suspicion of the consumer protection agency justified? What assumption is it necessary to make in this problem?

Here the sample size is small and we have, $n = 20$ $\bar{X} = 58$ $S = 8.3$ $df = 19$

a. For the 90% confidence interval, the tail area is 0.05 and therefore

$$58 - 1.729\frac{8.3}{\sqrt{20}} < \mu < 58 + 1.729\frac{8.3}{\sqrt{20}}$$
$$54.79 < \mu < 61.21$$

b. And for the 98% confidence interval, the tail area is 0.01 which gives us

$$58 - 2.539\frac{8.3}{\sqrt{20}} < \mu < 58 + 2.539\frac{8.3}{\sqrt{20}}$$
$$53.29 < \mu < 62.71$$

In the above two intervals, we see that with 90% and 98% confidence, the mean weight of the potato bags can plausibly be higher than 60 lb. Thus, the suspicion of the agency is not justified and there is not sufficient evidence to think that mean weight of the potato bags is below 60 lb. We return to this kind of problem regarding acceptance or rejection of a claim when we discuss test of hypotheses in the next chapter.

With regards to the necessary assumption in this problem, as discussed earlier, when the sample size is small, the above procedure is valid when the parent distribution from which the sample is taken has a normal distribution. Thus in this case, it is necessary to assume that the weight of the potato bags packaged by the company is a variable whose distribution can be described by the normal curve. Note also that in this case, as in the previous situation, the width of the confidence interval increases when the confidence level is higher.

Exercise 4

For the data on the amount of rainfall in cities of Florida that we saw earlier in this chapter, compute 90% and 95% confidence intervals for the mean amount of annual rainfall in all cities of Florida.

Section 5: Sample Size Determination

The question of sample size determination is a key design issue that any investigator faces. When a survey is planned, or a study is to be undertaken, one of the first questions that must be answered is how many individuals should be considered. The answer, will of course, depend on the level of accuracy and confidence the investigator expects from the study. The question, therefore is "how many items do we need to include in an experiment to estimate the population mean μ correct to within a bound (margin of error) B with a confidence coefficient of $1 - \alpha$?" As we said earlier in this chapter, the margin of error depends on two quantities, namely the sample size and the level of confidence. Thus if we fix the margin of error and the level of confidence, we can determine the sample size. Thus to answer the question about the required sample size, we set the desired error bound B equal to the margin of error, that is, $Z_{\frac{\alpha}{2}} \dfrac{\sigma}{\sqrt{n}} = B$. Solving this equation algebraically for n, we get

$$n = \left(\frac{\sigma \cdot Z_{\frac{\alpha}{2}}}{B} \right)^2$$

as the required sample size. We note that this expression will not necessarily lead to an integer. But, the sample size n has to be an integer value. Therefore, since the above expression is the minimum requirement for the stated specification, we generally round the result up to the nearest integer. Also, in practice, the value of the population standard deviation σ is often unknown and we usually replace it by some estimate of it based on previous experience or a pilot study. If nothing is available, we can get a rough estimate of population standard deviation by applying the empirical rule and dividing the sample range by 6 or more conservatively by 4.

■ Example 5

In the previous example, suppose the experiment is only a pilot study to estimate average heartbeats per minute for all male college athletes in the country. How many athletes should be included in the study in order to estimate the nationwide mean correct to 1.5 beats per minute with 99% confidence?

In Example 4, we found that for the sample selected in one college, the sample standard was 8.3 beats per minute. We use this value as an estimate of σ. Also, for this problem, the desired error bound B is said to be 1.5 beats per minute. Thus from the above formula we have

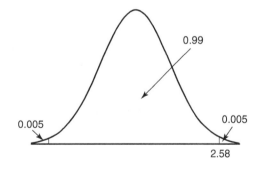

$$n = \left(\frac{8.3(2.58)}{1.5} \right)^2 = 203.80$$

Rounding this value up to the nearest integer, shows that we need $n \geq 204$, that is, at least 204 athletes have to be considered for the study to satisfy the requirements.

Exercise 5

Earlier in this chapter, we looked at a study to estimate the mean number of sick leave days for employees of a company. We first computed a 95% confidence interval and later 80% and 98% confidence intervals for the mean number of sick leave days for all employees. Suppose now that the company is not satisfied with the accuracy of the estimates and wishes to estimate the mean sick leave days more accurately with a possibly larger sample. How large of a sample should the company use if it wishes to estimate the true mean number of sick leave days correct to within one day with 95% confidence?

Hypotheses Testing

Section 1: Introduction

At the beginning of the last chapter, we mentioned that statistical inference consists mainly of two types of problems; namely estimation and hypotheses testing. In Chapter 7, we discussed the problem of estimation in a single sample. This chapter is devoted to the discussion of hypothesis testing, again in a single sample. In other words, the problem that we consider in this chapter pertains to a situation where we wish to use sample information to assess the validity of claims about the population. To understand the concept, we use an example to demonstrate and also to introduce the elements of a hypothesis testing problem.

Section 2: Elements of a Hypothesis Testing Problem

One of the parameters that pharmaceutical companies consider when developing oral drugs is the way the drug is absorbed is the bloodstream. For pain relievers, faster absorbing drugs seem to be preferred as it gives relief to the individual taking the drug in a shorter time.

■ Example 1: Application in Drug development

Suppose that a pharmaceutical company has developed a new drug which it claims to be better than all its competitors because, on average, it takes less time than all similar drugs to be absorbed in the bloodstream. Assume that the quickest similar drug currently available, takes an average of say 38.3 minutes to be absorbed into the bloodstream. Under what condition can we accept the claim of the company? We want to formulate this problem as a hypothesis testing problem.

The pharmaceutical company is claiming that their new drug takes on average less than 38.3 minutes to be absorbed. Thus if we denote by μ, the mean absorption time of the new drug, we can state the claim of the company as $\mu < 38.3$. Clearly this claim is not automatically accepted unless there is clear evidence in favor of it. Governments and regulatory agencies around the world are continuously faced with this problem. The Food and Drug Administration in the United States receives proposals for approval of new drugs from pharmaceutical companies on a regular basis and there are standard protocols for testing and examining the evidence before a drug is approved. So, in our example, the company claims that the mean time to absorption of the new drug is lower than 38.3 minutes. But, you, as the investigator of the company's claim, are not going to accept it. Rather, you believe in the contrary that the present drug is at best the same as what is currently available or may be even worse, unless otherwise proven to you. In other words you believe in a negating hypothesis of $\mu = 38.3$ or $\mu \geq 38.3$ unless there is evidence to suggest the invalidity of your belief.

Therefore we can see that we have two non-overlapping and exhausting hypotheses. For the sake of the formulation of the hypothesis testing problem, we name these two hypotheses. We call the hypothesis that you created to negate the hypothesis claimed by the company, the null hypothesis and denote it by H_0. Thus the null hypothesis is the hypothesis that states the *status quo*. The claimed hypothesis, on the other hand, is the research hypothesis and what in general is under investigation. We denote this hypothesis by H_a and call it the alternative hypothesis. Thus, we can see that any hypothesis testing problem consists of assessing the contrast between two non-overlapping hypotheses, i.e.,

$$\text{Null Hypothesis} - H_0: \mu \geq 38.3 \text{ or } \mu = 38.3$$

Versus

$$\text{Alternative Hypothesis} - H_a: \mu < 38.3$$

where we mention again that we are interested in verifying the validity of the research hypothesis. For this reason, you will find that in many hypotheses testing problems, the null hypothesis is stated as an equality. In the context of the above example, we are interested to know whether the new drug is superior to the best existing one. If the evidence leads to the fact that the new drug is not any better, then it really does not matter how much worse the new drug is compared to the existing one. Therefore we could equally and without loss of generality state the null hypothesis as $H_0: \mu = 38.3$.

Exercise 1

An electronics supplying company manufactures transistors to ship to other companies. The transistors are packaged in boxes of 300 transistors in each box. However, not all transistors are useable and each box can contain several defective transistors. The company claims that the mean number of defective transistors in each box is below 5. State the null and alternative hypotheses for this problem.

Null hypothesis:
Alternative hypothesis:

Exercise 2

The issue of elevated cancer incidence for residents living in proximity of a power plant has for long been a subject of controversy. In a random sample of 1900 individuals taken from a community living close to a power plant, 817 had been diagnosed as having some sort of cancer. According to National Cancer Institute (http://www.cancer.gov/about-cancer/what-is-cancer/statistics), approximately 39.6 percent of men and women are diagnosed with cancer at some point during their lifetimes (based on 2010–2012 data). Is there evidence of elevated cancer incidence rate in this community? State the null and alternative hypothesis for this problem.

    Null hypothesis:
    Alternative hypothesis:

Section 3: Sources and Types of Error

Clearly our decision to select one or the other hypothesis will depend on the evidence and information gathered through a sample. If we see sufficient evidence in favor of the claimed hypothesis, we reject the null and accept the alternative. On the contrary, if the sample evidence is insufficient to support the claim, we do not reject the null hypothesis and consequently cannot support the claim. Hence, based on the sample evidence, in any hypothesis testing problem, there are two possible decisions:

Decision 1 – Reject the null hypothesis in favor of the alternative hypothesis
Decision 2 – Fail to reject the null hypothesis and therefore do not accept the alternative

Notice that our decision is always stated in terms of the null hypothesis. The question here is whether we can make a wrong decision. Can we make an error in our decision-making process? Of course, even though our decision is based on sample evidence, it is possible to make a mistake. No decision process is ever perfect and, with a small chance, it is possible to commit error. Think about all the decisions that you made in the past. Can anyone ever say that all the decisions they have made have been the correct one? Look at the historical evidence. How many times have we heard that a decision in a court of law was later proved to be incorrect and it was reversed? How many times have we seen that a drug is recalled because of some safety problems? Sometimes these erroneous decisions can be very harmful and even fatal. In the late 1950s the thalidomide drug was widely introduced in markets all over the world for sleeplessness presumed to be completely safe for adults and children. In fact by 1960, thalidomide was marketed in 46 countries, with sales nearly matching those of aspirin. In the early 1960s, around the same time, the doctors started prescribing the drug to pregnant mothers to alleviate the morning sickness. Later, it was discovered that the babies from mothers who had used the drug during pregnancy had much higher rate of malformation incidence. The presumed 'harmless' drug had severely affected the normal tissue growth in babies. A decision maker makes a decision presumed to be correct at the instance of decision making based on available evidence. There is, of course, uncertainty associated with any decision. However, one can try to best make a decision by reducing uncertainties and minimizing the chance of error.

In any hypothesis testing problem, there are two sources of error. One source is when we find sufficient evidence in the sample to reject the null hypothesis and accept the alternative, while in reality the alternative hypothesis is the false one. We call this source of error *Type I*. The other source of error is the reverse situation, that is when we do not find sufficient evidence to reject the null and therefore we do not support the alternative, while in reality it is the alternative hypothesis that is the true one. We call this second type of error *Type II*.

Type I: Rejecting H_0 when H_0 is true
Type II: Not rejecting H_0 when H_0 is false

Note that, in general, since our decision is based on random sampling from the population, our decision is more likely to be a correct one. However, there is a chance that the random sample may not be a true representation of the population and therefore lead to a wrong decision. If you flip a coin 10 times, even if the coin is a perfectly unbiased one, there is a small probability that the coin lands on one side eight or more times. If this happens, you may erroneously declare the coin as biased. We denote the probability of Type I error by α and the probability of Type II error with β. Thus

α = Prob. (Type I Error)
β = Prob. (Type II Error)

It turns out that generally in hypotheses testing problems, the probability of Type I error is more serious than Type II.

■ Example 2: Application in Drug development—Continued

In the context of our demonstrating example regarding the drug development, Type I error occurs when we erroneously reject the null hypothesis and accept that the new drug as being superior when in reality it is really inferior. Type II error, on the other hand, occurs when we do not declare the new drug as being superior, when in reality it is a faster absorbing drug.

Thus in hypotheses testing problems, we fix the value of the probability of Type I error α at a low level, called the *significance level* of the test. We then try to use a testing procedure that leads to a low level of the probability of Type II error β. The quantity $1 - \beta$ is also called the *power* of the test. Thus in practice, depending on the nature of the problem, we select the value of α as a small number such as 0.10 or 0.05 and use a procedure that has relatively high power. Note that Type I and Type II errors are complementary events and that α and β do not add up to one. But, the lower we select the value of α, the higher is the chance of committing a Type II error and thus reducing the power of the test. The most common value of α in practice is 0.05, although depending on the problem, that could change. In some critical situations, one may choose the value of α as low of 0.01 or even lower. For example, in some medical and environmental problems when human lives may be at risk, the significance level may be selected to be very low. Similarly, in business when a large profit is at risk, the probability of Type I error is chosen at a low level.

Exercise 3

In the first exercise of this chapter, regarding the electronics supplying company, describe in words the two types of error. Which type of error appears to be more serious?

Type I Error:

Type II Error:

Section 4: Performing a Test of Hypotheses

We have therefore formulated a hypothesis testing problem as deciding and making a selection between two hypotheses, the null and the alternative hypotheses. Now, the question is how do we actually make that decision? As we demonstrated in our drug development example, the null and alternative hypotheses are formed in terms of population parameters. In reality, these hypotheses may concern one or more parameters. However, for our purpose we only consider problems that involve only one parameter. Also, among all the population parameters that we have encountered, here we only focus on hypotheses testing problems that concern the mean of the population μ.

As mentioned before, our decision regarding the validity of a hypothesis will depend on a random sample from the population. Clearly, when the decision is regarding the population mean μ, the most reasonable approach is to consider the value of the sample mean $\bar{X}$. If this value has a high probability of being close to the hypothesized value of the population mean, we intend to not reject the null hypothesis. However, we know that when the sample size is large, we have a powerful tool, namely the Central Limit Theorem, that gives us a good approximation for the distribution of the sample mean $\bar{X}$. For this reason, we consider two cases for testing the hypotheses regarding the population mean. First, we consider a situation where the sample size is large. We then go on to discuss the hypothesis testing problem regarding the population mean when the sample size is small. Remember that as a rule of thumb, whenever the sample size is 30 or more, we regard it as a large-size sample.

Large Sample Tests of Hypotheses for μ

Let us go back to our drug development example that we used to demonstrate the null and alternative hypotheses at the beginning of this chapter. Recall that drug developer claims that the time taken by the new drug to be absorbed in the bloodstream of the patient is less than all the similar drugs. We formulated the problem as deciding between H_0: $\mu = 38.3$ and H_a: $\mu < 38.3$. As mentioned above, since we wish to make a decision regarding the population mean, it makes sense to base our decision on the value of the mean of a random sample.

■ Example 3: Pharmaceutical Application

So, let us suppose that the drug is administered to a random sample of say $n = 50$ patients and for each patient the time for absorption of the drug is measured. Suppose that the mean and standard deviation of the sample come to $\bar{X} = 35.7$ and $S = 8.9$ respectively. Note that a sample size of 50 patients constitutes a large sample. So, the question now is whether the sample provides sufficient evidence for us to reject the null hypothesis and accept the alternative hypothesis. Clearly the sample mean of 35.7 is below the hypothesized value of 38.3. But, because of sample fluctuation and sampling variability we cannot conclude that the null hypothesis should be rejected and we should consider the sampling distribution of the sample mean. Let us for now assume that the null hypothesis is actually the true hypothesis. If the null hypothesis is true, then since the sample size is large, we can deduce by the Central Limit Theorem that the sampling distribution of the sample mean $\bar{X}$ is normal with mean $\mu_{\bar{X}} = 38.3$ and standard deviation of $\sigma_{\bar{X}} = \dfrac{\sigma}{\sqrt{n}}$, where σ is the population standard deviation. Therefore

$$z = \frac{\bar{X} - \mu_{\bar{X}}}{\sigma / \sqrt{n}}$$

has a standard normal distribution. Clearly, in order to compute the value of the z statistic for the problem at hand, we need to have the value of σ, which is unknown. Thus we use its point estimate, which is the sample standard deviation S. Note that, as we saw in Chapter 7, since the sample size is large, we can argue that substituting the sample standard deviation in place of the population standard deviation will not alter the distribution of the test statistic. Thus the value of the test statistic z is

$$z = \frac{35.7 - 38.3}{8.9 / \sqrt{50}} = -2.07$$

Now, the question is how likely is it to observe a Z value of -2.07 or lower by chance. To answer this question, we compare this value with the tail areas of the standard normal curve. Suppose for the purpose of this example, we set the significance level of the test as $\alpha = 0.05$. This means that we are allowing a maximum probability of 0.05 for Type I error, i.e., rejecting the null hypothesis by mistake. Using the normal distribution chart or the invNorm function on the calculator, we find that 0.05 corresponds to a Z value of -1.645 on the left tail of the normal curve. Now, we see that the observed value of -2.07 is even more extreme than this so called *critical value*. In other words, the chance of observing a Z value -2.07 or lower by chance is even less than the nominal 5%. This indicates that our assumption about the truth of the null hypothesis is false and therefore we should decide on rejecting the null hypothesis. In statistical terminology, we say that the *test is significant* and there is sufficient evidence to substantiate the alternative hypothesis. Therefore our conclusion is that at the 5% level of significance, the claim of the pharmaceutical company regarding the superiority of their new product is supported.

Exercise 4

In the previous exercise regarding the electronics supplying company and the mean number of defective transistors per box, suppose we randomly sample 43 boxes and count the number of defectives in each box. After calculating the mean and standard deviation of the number of defectives, we find that the sample mean was 4.42 with a standard deviation of 1.89. Calculate the test statistic and decide whether to reject the null hypothesis when $\alpha = 0.05$ and when $\alpha = 0.01$. Give an interpretation of your results.

Test Statistic:

Decision Rule for $\alpha = 0.05$:

Decision:

Interpretation:

Decision Rule for $\alpha = 0.01$:

Decision:

Interpretation:

Section 5: One-Sided and Two-Sided Tests

In the above example, note that the claim of the pharmaceutical company about the mean time for absorption of the new drug and consequently the alternative hypothesis were such that the critical region fell on the left side of the tail of the normal distribution. Clearly, there are many instances in practice that we wish to verify the validity of a claim about the mean of a population being greater than a certain value. For example, a manufacturer of appliances might claim that the mean life of their brand of refrigerator exceeds a certain number. In that case, the critical region of the test will fall on the right tail of normal distribution, but the procedure for decision making is similar to what we discussed above. Similarly, there may be situations when the alternative hypothesis is to verify whether the mean of the distribution is not equal to a certain value. In that case, the critical region falls on both sides of the normal distribution. While we call the two cases where the critical region is on one side of the normal curve, *one sided*, this latter case with the critical region on both tails of the curve is called a *two-sided* test. Once again, the procedure is similar, but in this case, we split the significance level α between the two tails and determine the critical values in such a way that each tail has an area of $\alpha/2$. We can summarize the procedure for *large sample test of hypotheses about the population mean* as follows:

In testing

$$H_0: \mu = \mu_0 \qquad \text{versus} \qquad H_a: \mu \leq \mu_a$$

We calculate the test statistic $\quad Z_0 = \dfrac{\bar{X} - \mu_0}{\sigma/\sqrt{n}}$

If the significance level of the test is set at α, the null hypothesis H_0 is rejected when $Z_0 \leq -Z_\alpha$, where Z_α is the critical value determined so that the right tail area of the normal curve is α.

In testing

$$H_0: \mu = \mu_0 \qquad \text{versus} \qquad H_a: \mu \geq \mu_a$$

the test statistic $\quad Z_0 = \dfrac{\bar{X} - \mu_0}{\sigma/\sqrt{n}}$

is calculated and at the α level of significance the null hypothesis H_0 is rejected when $Z_0 \leq -Z_\alpha$ where, as before, Z_α is the critical value determined so that the right tail area of the normal curve is α.

The above two cases are for the one-sided tests. For the two-sided test, the following procedure is used:

In testing

$$H_0: \mu = \mu_0 \qquad \text{versus} \qquad H_a: \mu \neq \mu_a$$

the same test statistic $\quad Z_0 = \dfrac{\bar{X} - \mu_0}{\sigma/\sqrt{n}}$

is calculated and null hypothesis is rejected when either $Z_0 \leq -Z_{\alpha/2}$ or $Z_0 \geq -Z_{\alpha/2}$. Note that in this case we have a two-sided rejection region, where the significance level α is divided equally between the two tails.

■ Example 4: Lifetime of Tires

A tire manufacturer claims that the average life of their new line of production exceeds 50,000 miles. To check this claim, we randomly sample 36 tires and through some accelerated testing device, measure the life of each tire. We find that the sample mean and standard deviation are respectively 54,365 miles and 8,435 miles. Do the data support the manufacturer's claim at $\alpha = 0.05$?

Unlike our first example, we see that in this case, we wish to verify the validity of a hypothesis that the mean of the distribution of the lifetime of tires μ exceeds a certain specified value. Thus the null and alternative hypotheses can be stated as follows:

$$H_0: \mu = 50,000 \qquad \text{versus} \qquad H_a: \mu > 50,000$$

Hence we have a one-sided test and the critical region this time falls on the right side of the normal distribution. Since α is specified at 0.05, we look for a value on the normal curve for which the area to the right of that value is 0.05. Once again, using a normal distribution chart or the invNorm function on our calculator, we find that this critical value is 1.645. Thus our decision rule is

Decision Rule: Reject the null hypothesis if the value of the test statistic exceeds 1.645

Here, we have $n = 36$, $\bar{X} = 54,365$, and $S = 8,435$. Thus the value of the test statistic is calculated as

$$Z_0 = \frac{54365 - 50000}{8435 / \sqrt{36}} = 3.10$$

and we see that the value of the test statistic falls in the rejection region. Thus the test is significant and our decision is to reject the null hypothesis. This, in turn, means that there is sufficient evidence to substantiate the manufacturer's claim.

■ Example 5: Fruit Juice Bottle Filling

A fruit juice filling machine is set to fill each bottle with 250 mL of the juice. However, there is some fluctuation and the content of the bottles is not exactly 250 mL. The quality control department of the fruit juice company is interested to know if it can be stated that the average content is 250 mL. They randomly sample 80 bottles and accurately measure the content of each bottle. They find that the average for the 80 sampled bottles is 248.79 mL with a standard deviation of 9.23 mL. We wish to test the hypothesis that the mean content is actually 250 mL at $\alpha = 0.01$.

Denoting the mean fruit juice content of all bottles filled by this machine by μ, here, the null and alternative hypotheses can be stated as

$$H_0: \mu = 250 \qquad \text{versus} \qquad H_a: \mu \neq 250$$

and so we have a two-sided test. Thus, our critical region is also two-sided falling on both tails of the normal curve. Dividing the given value of α equally between the two tails, we find that the area for each tail is 0.005. Using the corresponding values of the standard normal scores, our decision rule is

Decision Rule: Reject the null hypothesis if the value of the test statistic is either less than −2.575 or higher than 2.575.

Now, the value of the test statistic is calculated as:

$$Z_0 = \frac{248.79 - 250}{9.23 / \sqrt{80}} = 1.17$$

which indicates that the test is not significant and the value of the test statistic does not fall in the rejection region. Therefore we fail to reject the null hypothesis and accept the fact that the mean content of all the bottles filled by this machine is 250 mL.

✔ Steps for Hypotheses testing

We see that in order to perform a test of hypothesis, the following steps are taken:

1. State null and alternative hypotheses.
2. Decide whether the test is one-sided or two-sided.
3. Based on the given significance level, determine the critical (rejection) region.
4. Compute value of the test statistic z_0.
5. Decide whether value of test statistics falls in the critical region.
6. Decide whether the test is significant (reject null hypothesis H_0) or not significant (do not reject null hypothesis H_0).
7. Provide an interpretation of results.

Exercise 5

The Center for Disease Control and Prevention reported in 2011 that the average Systolic Blood Pressure (SBP) of all adults is estimated to 122 (http://www.cdc.gov/).

In a certain geographic area, it is thought that the average SBP exceeds that of the general population. A random sample of 200 adults resulted in an average SBP of 128 with a standard deviation of 22. Using $\alpha = 0.05$ and $\alpha = 0.01$, follow the steps and test the relevant hypotheses.

Exercise 6

The per capita income for the United States in the year 2015 was reported by the World Bank to be $56,084 (http://en.wikipedia.org).

The governor of a state is interested to test whether or not the average per capita income in his/her state matches the rest of the country. A random sample of 55 citizens gave an average per capita income of $53,988 with a standard deviation of $5,107. Test the relevant hypotheses and draw conclusions with $\alpha = 0.1$ and $\alpha = 0.02$.

Section 6: Observed Significance Level (*P*-Values)

In the last section, we learned that in a hypothesis testing problem, there are two sources of error. Owing to the fact that we generally believe that Type I error can lead to more serious consequences, in practice, we fix the maximum Type I error that we are willing to tolerate. In other words we fix what we called the α-level of the test and our decision is based on this value. Clearly, this choice of α plays an important role in the whole decision making process. In fact in some cases, even small changes in the value α can completely reverse the decision. Let's take a look at an example.

■ Example 6: Pharmaceutical Application

In Example 3, for testing the null hypothesis H_0: $\mu = 38.3$ against H_a: $\mu < 38.3$ regarding the mean time for absorption of a drug, we found that value of the test statistic was $Z_0 = -2.07$. Since the value of α was set at $\alpha = 0.05$, the critical value was calculated as -1.645, which in turn led to significance of the test and rejection of the null hypothesis. This meant that sufficient evidence was found in favor of the new drug. Suppose now, that the agency examining the drug, takes a more stringent approach and believes that the maximum Type I error that can be tolerated in this test, is $\alpha = 0.01$. With this new value of α, clearly the critical value and consequently, the decision rule will change. Indeed, using a normal distribution chart or invNorm(0.01) in our calculator, we find that the new critical value is -2.33 and so, our decision rule becomes:

Decision Rule: Reject the null hypothesis if $Z_0 < -2.33$.

With the calculated value of the test statistic at -2.07, we see that our decision is completely reversed and the null hypothesis is not rejected, which implies that we do not accept the new drug as being better.

We see therefore the choice of the significance level α is a very crucial one and naturally in many cases hard to make. In fact, the statistician may prefer to leave the decision for the value of α to the expert. Rather than fixing the value of α, and making a definite decision, it is possible to report the *Observed Significance Level* or what is commonly known as the *P-value* for the test. Loosely, the *P*-value of a test describes how much support there is in the data in favor of the null hypothesis and so when this support is low enough, we tend to reject the null hypothesis. Formally, the *P*-value is the probability of observing a value of the test statistic as extreme or more extreme than the one observed.

■ Example 6: Pharmaceutical Application (Continued)

Hence we can see that the *P*-value for the test in Example 4 is the probability of observing a normal score that is -2.07 or lower and thus the *P*-value can be calculated as

P-value $= P(Z \leq -2.07) = 0.0192$

Exercise 7

For the exercise regarding the mean number of defective transistors, compute the *P*-value of the test and interpret it.

Section 7: A Note Regarding *P*-Values

Note that in Example 6, the *P*-value of 0.0192 is lower than 0.05, but higher than 0.01. Recall that the null hypothesis was rejected at $\alpha = 0.05$ and not rejected at $\alpha = 0.01$. Thus rather than making a decision, the *P*-value of the test is determined and the decision is left to the expert. If the expert decides that an α-level of 0.05 is adequate, the null hypothesis is rejected. If, on the other hand, the expert decides that the α-level should be 0.01, then the null hypothesis is not rejected. In fact from the *P*-value for this test, we infer that if the α-level is 0.04, or 0.03 or even 0.02, still the null hypothesis is rejected. Thus, if we are given an α-level, as long as the *P*-value is lower than the α-level, the test can be considered significant and the null hypothesis is rejected.

Summary of *P*-Value Calculation

In a one-tailed test,

For testing $H_0: \mu = \mu_0$ against $H_a: \mu < \mu_0$ *P*-value $= P(Z \leq Z_0)$

$H_0: \mu = \mu_0$ against $H_a: \mu < \mu_0$ *P*-value $= P(Z \geq Z_0)$

and for a two-tailed test,

For testing $H_0: \mu = \mu_0$ against $H_a: \mu \neq \mu_0$ *P*-value $= 2P(Z \geq |Z_0|)$

where in the above, as before, $Z_0 = \dfrac{\bar{X} - \mu_0}{\sigma / \sqrt{n}}$ is the test statistic. Note, once again, that if the α-level of the test is given, the null hypothesis is rejected when *P*-value $< \alpha$-level.

■ Example 7: Lifetime of Tires

In Example 4, we were interested to test whether the mean life of a specific type of tire was higher than 50,000 miles. In that problem, we found that the value of the test statistic was calculated as $Z_0 = 3.10$ and therefore the *P*-value can be calculated as

P-value $= P(Z \geq 3.10) = 0.00097$

Note that in this case the *P*-value is very small. This indicates that no matter how stringent the test is and how small the value of α is selected, we are still likely to reject the null hypothesis. This, in turn, means that there is a strong evidence in favor of the alternative hypothesis and we can support the claim that the mean life of the tires exceeds 50,000 miles.

■ Example 8: Fruit Juice Bottle Filling

In example 5, we had a two-tailed test to examine whether the mean amount of fruit juice filled by a machine can be considered to be 250 ml. We found that the value of the test statistic was $Z_0 = 1.170$. Thus, because we have a two-sided alternative, the *P*-value is calculated as

P-value $= 2P(Z \geq 1.17) = 2(0.121) = 0.242$

Unlike the previous example, this *P*-value is quite high and there is weak evidence to reject the null hypothesis and believe that the mean content of the bottles is different from 250 ml. Indeed even if the α-level is set as high as 0.2, we still fail to reject the null hypothesis.

Exercise 8

For the two exercises regarding the blood pressure of adults and the mean per capita income, calculate the P-value in each problem and interpret it.

Section 8: Small Sample Test of Hypothesis for μ

As discussed in Chapter 7, there are many instances, in practice, that large samples are not feasible. We gave some examples in Chapter 7 where we saw that due to cost, duration, or morality, a large sample size was not possible. We learned how to find confidence intervals for the population mean when only a small sample sizes are available. Here we discuss the problem of hypothesis testing in those situations.

Suppose we wish to test the null hypothesis H_0: $\mu = \mu_0$ against a one-sided or a two-sided alternative hypothesis based on a random sample of size n (small) from a population with mean μ and standard deviation σ. Note that once again, because of the small sample size, we cannot apply the Central Limit Theorem and we focus our attention only to situations where the distribution of the population from which the sample is taken is normal. Then, under the null hypothesis, we know that the statistic

$$T = \frac{\bar{X} - \mu_0}{S/\sqrt{n}}$$

has a t-distribution with $(n-1)$ degrees of freedom. Therefore, comparing to the critical values of the t-distribution we can state the recipe for small sample hypothesis testing for μ as follows:

To test:

$$H_0: \mu = \mu_o \text{ against } H_a: \mu < \mu_o$$

Compute:

$$T = \frac{\bar{X} - \mu_0}{S/\sqrt{n}}$$

Reject H_0 if $T < -\text{t-crit}$

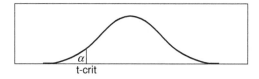

To test:

$$H_0: \mu = \mu_o \text{ against } H_a: \mu > \mu_o$$

Compute:

$$T = \frac{\bar{X} - \mu_0}{S/\sqrt{n}}$$

Reject H_0 if $T < -\text{t-crit}$

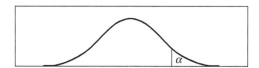

To test:

$$H_0 = \mu = \mu_o \text{ against } H_a = \mu \neq \mu_o$$

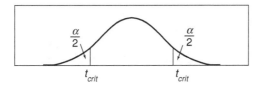

Compute $T = \dfrac{\bar{X} - \mu_0}{S \big/ \sqrt{n}}$ and Reject H_0 if $\text{T} < -t_{crit}$ or $\text{T} > t_{crit}$

■ Example 9: Pharmaceutical Application

In order to assess toxicity of a chemical compound, a dose of the chemical is injected to 20 rats, and time to an adverse effect (e.g., tumor development) is observed. It is found that the average time to effect is 56.8 days with a standard deviation of 10.3 days. Is there evidence to suggest that rats exposed to this chemical show an adverse effect in less than 60 days? Test using $\alpha = 0.05$. What assumption are we making in this problem?

$$H_0: \mu = 60 \qquad n = 20 \quad \bar{x} = 56.8$$
$$H_a: \mu < 60 \qquad \qquad \sigma = 10.3$$

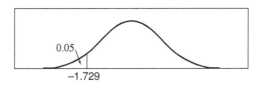

Decision Rule: Reject H_0 if $T < -1.729$

$$T = \frac{56.8 - 60}{10.3 \big/ \sqrt{20}} = -1.39$$

Clearly -1.39 does not fall in the rejection region and we fail to reject H_0. The test is not significant and there is insufficient evidence to show that it takes less than 60 days to develop an adverse effect for rats exposed to the chemical. Of course, the assumption of normality is necessary for the validity of the test. Thus we assume that time for the development of an adverse effect is a variable with a normal distribution.

Exercise 9

An online pizza delivery chain advertises that once your order is in, the pizza is delivered to your house in less than 25 minutes. Because of your past experience, you are a little dubious about this claim and ask six friends to exactly time the delivery after they order online. Here are the results (in minutes):

$$27.6 \quad 24.3 \quad 23.9 \quad 26.2 \quad 25.6 \quad 25.9$$

Based on this data, is the claim of the chain justified? State the null and alternative hypotheses and perform the test with $\alpha = 0.05$.

Exercise 10

A machine is set to fill bottles with exactly 8 ounces of fluid. Of course due to machine variability it is understood that although there are some fluctuations in the amount of fluid dispensed from bottle to bottle, the average is still 8 ounces. However, recently the quality engineer has become concerned that the machine is no longer dispensing an average of 8 ounces. She randomly samples 25 bottles from the line and precisely measures the fluid content. She finds that the average for the sample is 7.38 ounces with a standard deviation of 0.82 ounces. Is there evidence to suggest that the mean content of the bottles filled with this machine is different from 8 ounces? Test using $\alpha = 0.01$.

Comparing Two Populations

Section 1: Introduction

The last two chapters were devoted to a discussion of a very important branch of statistics, namely inference. We learned that inferential problems are generally of two types, either estimation or hypothesis testing. In both chapters we were dealing with a single sample from a single population and the inference was about the mean of the population. In practice, however, there are many instances where we are interested to compare two populations. For example, if a company has two sites for production, it would be interested to know whether there is a difference in the mean productivity of the two plants. A public health official might be interested to know whether there is a difference in the mean age of women and men seeking flu vaccination. A standard design in the medical and biological studies to examine the effect of a treatment is to use a so-called case-control design. In such studies, individuals are randomly assigned to one of the two groups. One group is used for treatment, e.g., is prescribed with the drug and the other group is used as control, e.g., is given the placebo. The results of the two groups are then compared to see if the treatment has any effect.

In this chapter, therefore, we study the inferential problems for comparing two populations. We discuss both estimation and hypothesis testing and assume that samples are available from both populations. Once again, because the methodology varies for large samples and small samples owing to the fact that in large samples we can apply the Central Limit Theorem, we consider large sample and small sample inferential problems separately. Also, analogous to Chapters 7 and 8, we only consider inferential problems for the population mean.

It is to be noted that, in practice, there are circumstances where we wish to compare more than two populations. For example, one may be interested to know if a difference exists between the performances of four different brands of gasoline on a car or five similar drugs for treating certain problem. There are statistical procedures that can be used for comparison of more than two populations, but the discussion of these procedures is beyond the scope of this manuscript and interested students can take more advanced courses in statistics to learn about those procedures.

Section 2: Inferences for Two Population Means in Large Samples

Suppose that a random sample of size n_1 (large) is given from a population with mean μ_1 and standard deviation σ_1. Also suppose that a random sample of size n_2 (large) is given, independently of the first sample, from another population with mean μ_2 and standard deviation σ_2. We are interested to make inferences about the difference in the population means $\mu_1 - \mu_2$. Let us suppose that $\bar{X}_1$ and S_1 are the mean and standard deviation of the first sample and similarly, $\bar{X}_2$ and S_2 are the mean and standard deviation of the second sample, respectively. We understand from Chapter 7 that the sample mean provides an adequate point estimate for the population mean. Thus we can deduce that $\bar{X}_1 - \bar{X}_2$ also provides an adequate point estimate for $\mu_1 - \mu_2$. Since both $\bar{X}_1$ and $\bar{X}_2$ are calculated from random samples, clearly the value of $\bar{X}_1 - \bar{X}_2$ varies from one set of samples to another. So, the question that naturally arises here is what is the sampling distribution of $\bar{X}_1 - \bar{X}_2$?

Because n_1 and n_2 are both large, it turns out that by an application of the Central Limit Theorem, the distribution of $\bar{X}_1 - \bar{X}_2$ is normal with mean

$$\mu_{\bar{X}_1 - \bar{X}_2} = \mu_1 - \mu_2$$

and standard deviation

$$\sigma_{\bar{X}_1 - \bar{X}_2} = \sqrt{\frac{\sigma_1^2}{n_1} + \frac{\sigma_2^2}{n_2}}$$

Please note that in the above formula for the standard deviation of $\bar{X}_1 - \bar{X}_2$, the variances of $\bar{X}_1$ and $\bar{X}_2$ are added in the expression under the square root sign. Now, by standardizing $\bar{X}_1 - \bar{X}_2$, we can assay that

$$Z = \frac{(\bar{X}_1 - \bar{X}_2) - (\mu_1 - \mu_2)}{\sqrt{\dfrac{\sigma_1^2}{n_1} + \dfrac{\sigma_2^2}{n_2}}} \tag{1}$$

has a standard normal distribution. Hence, we can set up inferential procedures using the tail values of the normal distribution similar to Chapters 7 and 8.

A. Large Samples Confidence Interval for $\mu_1 - \mu_2$

If it is desired to construct a confidence interval for $\mu_1 - \mu_2$ with confidence level of $1-\alpha$, so that the total tail area is α, we have

$$P\left(-Z_{\frac{\alpha}{2}} < Z < Z_{\frac{\alpha}{2}}\right) = 1 - \alpha.$$

Substituting for Z from (1) above and solving for $\mu_1 - \mu_2$, we get

$$(\bar{X}_1 - \bar{X}_2) - Z_{\alpha/2}\sqrt{\frac{\sigma_1^2}{n_1} + \frac{\sigma_2^2}{n_2}} < \mu_1 - \mu_2 < (\bar{X}_1 - \bar{X}_2) + Z_{(\alpha/2)}\sqrt{\frac{\sigma_1^2}{n_1} + \frac{\sigma_2^2}{n_2}} \tag{2}$$

which produces the desired interval. Note the similarity of the above formula with what we derived for the large sample confidence interval for the mean of a single population in Chapter 7. Both formulas have the same structure, but here we have a natural extension for two populations.

B. Large Samples Test of Hypotheses for $\mu_1 - \mu_2$

Here, the null and alternative hypotheses are stated in terms of $\mu_1 - \mu_2$. Specifically, suppose we wish to test the null hypothesis that the difference in the means is some specified value D_0, that is $H_0 : \mu_1 - \mu_2 = D_0$ against a one-sided alternative $H_a : \mu_1 - \mu_2 < D_0$. In most applications D_0 is zero as we are interested to know whether the means of the two populations are equivalent. Using (1) above, under the null hypothesis, the test statistic is

$$Z_0 = \frac{(\bar{X}_1 - \bar{X}_2) - D_0}{\sqrt{\dfrac{\sigma_1^2}{n_1} + \dfrac{\sigma_2^2}{n_2}}}$$

And analogous to what we discussed in Chapter 8, we reject the null hypothesis if the value of the test statistic is more extreme than the tail area of the normal distribution. Specifically, if $Z_0 < -Z_\alpha$. Similarly to test the null hypothesis

$$H_0: \mu_1 - \mu_2 = D_0$$

versus

$$H_a: \mu_1 - \mu_2 > D_0$$

we compute Z_0 and reject H_0 if $Z_0 > Z_\alpha$.

Finally, for a two-sided alternative, to test the null hypothesis

$$H_0: \mu_1 - \mu_2 = D_0$$

against

$$H_a: \mu_1 - \mu_2 \neq D_0$$

we have a two-sided rejection region and reject the null hypothesis if $Z_0 < -Z_{\alpha/2}$ or $Z_0 > Z_{\alpha/2}$.

Before we embark on some examples, we mention here, that once again in practice the population standard deviations σ_1 and σ_2 may be unknown in which case we replace them by their respective point estimates, the sample standard deviations S_1 and S_2. Also, similar to what we discussed in Chapter 8, here we can calculate a P-value for the test and it would have the same interpretation.

Example 1: Comparing the rate of heartbeats for men and women

In order to compare the mean heart rate of female and male college athletes, random samples of 36 female and 40 male college athletes were taken. We found that for female athletes, the mean number of heart beats per minute was 64 with a standard deviation of 8, while for male athletes the mean was 58 with a standard deviation of 6.

  a. Find a 95% confidence interval for difference in mean heart rates of all college female and male athletes.
  b. Is there evidence to suggest that the mean number of heart beats per minute for female college athletes is different from male athletes? Test using $\alpha = 0.05$.
  c. Compute the P-value of the test

SOLUTION

  a. Let μ_1 be the mean heart rate of all male athletes and μ_2 be the mean heart rate of all female athletes. Then, from the information given in the problem, we have

$$n_1 = 36 \quad \bar{X}_1 = 64 \quad S_1 = 8$$
$$n_2 = 40 \quad \bar{X}_2 = 58 \quad S_2 = 6$$

And since from the standard normal distribution $z_{0.025} = 1.96$ from (2) above, we have

$$(64-58)-1.96\sqrt{\frac{8^2}{26}+\frac{6^2}{40}} < \mu_1 - \mu_2 < (64-58)+1.96\sqrt{\frac{8^2}{26}+\frac{6^2}{40}}$$

$$2.79 < \mu_1 - \mu_2 < 9.21$$

b. Here, the test of hypotheses has a two-sided alternative and therefore the null and alternative hypotheses are respectively given by

$$H_0: \mu_1 - \mu_2 = 0$$
$$H_a: \mu_1 - \mu_2 \neq 0$$

And thus for a two-sided rejection region with $\alpha = 0.05$, the rejection rule is:

Reject H_0 if $z_0 < -1.96$ or $z_0 > 1.96$

Now, the value of the test statistic is:

$$z_0 = \frac{(64-58)-0}{\sqrt{\frac{8^2}{26}+\frac{6^2}{40}}} = 3.67$$

which is greater than 1.96 and so the test is significant. We reject H_0 and deduce that there is insufficient evidence to suggest that the mean number of heartbeats per minute is different in male athletes and female athletes.

c. The *P*-value of the test is calculated as

$$P\text{-value} = P(z < 3.67) + P(z > 3.76) = 0 + 0 = 0$$

which clearly shows that the test is highly significant and there is strong evidence to support the alternative hypothesis.

Exercise 1

The gender gap in salary is the subject of many discussions and controversies. Suppose that in one study, we wish to investigate the difference in salaries of men and women with professional jobs who at least have a college degree. We randomly select 110 men and 100 women professionals with college degrees and determine their annual salaries. We find that the average salaries for men are $62,850 with a standard deviation of $4,620 while for women the average is $59,460 and the standard deviation is $3,150.

a. Construct 90% and 98% confidence intervals for the mean difference in salaries of men and women with at least a college degree.

b. Is there evidence to think that mean salaries of men and women with at least a college degree are different? Test using $\alpha = 0.01$.

c. What is the P-value of the test in part b?

d. It is thought that the mean salary of men with at least a college degree is higher than the women's by more than $3,000 per year. Is there evidence to support such a claim at $\alpha = 0.05$ level of significance? What is the P-value for such a test?

e. Do you think the results of this study are reliable? What other factors might affect the salaries that are not accounted for in this study?

Section 3: Inferences for Two Population Means in Small Samples

Suppose once again that we have two independent samples from two populations and we are interested to make inferences about the difference in population means $\mu_1 - \mu_2$, but now both sample sizes n_1 and n_2 are small. Clearly, the Central Limit Theorem does not apply. Recall that in Chapter 7 when we were discussing the interval estimation of the population mean in small samples, we focused only on populations whose distribution could be assumed to be normal. Similar to the situation in the one-sample case, we restrict this development to normal populations only. We assume, therefore, that the parent populations from which the samples are drawn are both normal. Here, we also make another assumption. We further assume that the two populations have the same variability that is to say that variances of the two populations σ_1^2 and σ_2^2 are equal. This means that even though the two normal curves that describe the shape of the population distributions may be centered at different locations, they have the same shapes and one is just a shifted copy of the other one. We denote the common variance by σ^2. Then the sampling distribution of $\bar{X}_1 - \bar{X}_2$ is again normal with $\mu_{\bar{X}_1 - \bar{X}_2} = \mu_1 - \mu_2$ and

$$\sigma_{\bar{X}_1 - \bar{X}_2} = \sqrt{\frac{\sigma^2}{n_1} + \frac{\sigma^2}{n_2}} = \sigma\sqrt{\frac{1}{n_1} + \frac{1}{n_2}}. \text{ Thus}$$

$$Z = \frac{(\bar{x}_1 - \bar{x}_2) - (\mu_1 - \mu_2)}{\sigma \cdot \sqrt{\frac{1}{n_1} + \frac{1}{n_2}}}$$

has a standard normal distribution and we can base our inferences on this statistic. But in the above expression, σ is unknown and has to be estimated. To estimate the common standard deviation σ, we pool both samples and use a combined estimate for σ. Specifically, we define

$$S_p^2 = \frac{(n_1 - 1)S_1^2 + (n_2 - 1)S_2^2}{(n_{1-1}) + (n_2 - 1)}$$

as the pooled estimate of the variance σ^2 whose square root S_p provides an estimate of σ. Now, because n_1 and n_2 are not large, replacing σ by S_p in the above expression will no longer preserve normality and the statistic

$$T = \frac{(\bar{X}_1 - \bar{X}_2) - (\mu_1 - \mu_2)}{S_p \cdot \sqrt{\frac{1}{n_1} + \frac{1}{n_2}}}$$

has a t-distribution with $n_1 + n_2 - 2$ degrees of freedom. This can become a basis for inferences about $\mu_1 - \mu_2$ in small samples.

A. Small Samples Confidence Interval for $\mu_1 - \mu_2$

Thus a confidence interval for $\mu_1 - \mu_2$ with confidence coefficient of $1 - \alpha$ is given by

$$(\bar{X}_1 - \bar{X}_2) - t_{a/2} . S_p . \sqrt{\frac{1}{n_1} + \frac{1}{n_2}} < \mu_1 - \mu_2 < (\bar{X}_1 - \bar{X}_2) + t_{a/2} . S_p . \sqrt{\frac{1}{n_1} + \frac{1}{n_2}}$$

where $t_{a/2}$ is the critical value determined from the t-distribution with $n_1 + n_2 - 2$ degrees of freedom in such a way that the tail area is $a/2$.

B. Small Samples Test of Hypotheses for $\mu_1 - \mu_2$

Similarly to test the null hypothesis of $H_0: \mu_1 - \mu_2 = D_0$ against a one-sided or a two-sided alternative, the test statistic

$$T_0 = \frac{(\bar{X}_1 - \bar{X}_2) - D_0}{S_p \cdot \sqrt{\dfrac{1}{n_1} + \dfrac{1}{n_2}}}$$

may be used in the usual manner. Thus if the alternative hypothesis is $H_a: \mu_1 - \mu_2 < D_0$, then the null hypothesis is rejected when $T_0 < -t_\alpha$. If the alternative hypothesis is $H_a: \mu_1 - \mu_2 < D_0$, then the null hypothesis is rejected for $T_0 > t_\alpha$ and finally if we have a two-sided alternative $H_a: \mu_1 - \mu_2 \neq D_0$, then we reject H_0 when $T_0 < -t_{\alpha/2}$ or $T_0 > -t_{\alpha/2}$. Note that the critical values of the t-distribution t_α and $t_{\alpha/2}$ are based on $n_1 + n_2 - 2$ degrees of freedom.

■ Example 2: Comparing the Lifetime of Batteries

We wish to compare the lifetimes of two brands of similar batteries. A random sample of 10 batteries of brand 1 and a random sample of 7 batteries of brand 2 are selected and the life of each battery is determined. It is found that for brand 1, the mean life is 4.6 hours with a standard deviation of 0.65 hour while for brand 2, the average life is 4.1 hours with a standard deviation of 0.48 hour.

a. Estimate the difference in the mean life of all batteries of the two brands 1 and 2 comparing the rate of heartbeats for men and women using a 95% confidence interval.
b. Is there evidence to suggest that on average brand 1 batteries last longer than brand 2 batteries? Test using $\alpha = 0.05$.
c. What assumption is necessary to be made in this problem?

SET UP

We have

$$n_1 = 10 \qquad n_2 = 7$$
$$\bar{X}_1 = 4.6 \qquad \bar{X}_2 = 4.1$$
$$S_1 = 0.65 \qquad S_2 = 0.48$$

Therefore, the pooled estimate of the variance is given by

$$S_p^2 = \frac{9(0.65)^2 + 6(0.48)^2}{9 + 6} = 0.345$$

and thus

$$S_p = 0.588.$$

Note that the number of degrees of freedom for the t-statistic is $10 + 7 - 2 = 15$.

SOLUTION

a. Using a critical value of the t-statistic with 15 degrees of freedom at $\alpha = 0.025$, we have

$$(4.6 - 4.1) - (2 \cdot 131)(\cdot 588)\sqrt{\frac{1}{10} + \frac{1}{7}} < \mu_1 - \mu_1 < (4.6 - 4.1) + (2 \cdot 131)(\cdot 588)\sqrt{\frac{1}{10} + \frac{1}{7}}$$
$$-0.1175 < \mu_1 - \mu_2 < 1.1175$$

b. We wish to test $H_0: \mu_1 - \mu_2 = 0$ against $H_\alpha: \mu_1 - \mu_2 > 0$. We have

$$T = \frac{(4.6 - 4.1) - 0}{.588\sqrt{\dfrac{1}{10} + \dfrac{1}{7}}} = 1.725$$

Since the critical value of the t-statistic with 15 degrees of freedom and tail area of $\alpha = 0.05$ is 1.753 and $1.725 < 1.753$, we cannot reject the null hypothesis and conclude that there is insufficient evidence to suggest that on average, batteries of brand 1 last longer than batteries of brand 2.

c. It is necessary to assume that the two samples are from populations whose distributions are normal. This means that we assume that the lifetime of batteries from each brand is a variable whose distribution can be characterized by the normal curve.

Exercise 2

To compare the average mileage of SUVs manufactured by two competitive companies, a random sample of 5 is taken from the first company and the mileage per gallon is measured for each car:

$$18.4 \quad 20.1 \quad 16.9 \quad 17.8 \quad 19.5$$

In a similar experiment, a random sample of 7 SUVs is taken from the second company and mileage per gallon is measured for each one:

$$16.8 \quad 19.2 \quad 15.9 \quad 18.9 \quad 19.1 \quad 17.3 \quad 16.5$$

a. Compute a 95% confidence interval for the mean mileage difference for the SUVs manufactured by the two companies

b. Is there evidence to indicate that there is a discernible difference in the mean mileage performances? Test using $\alpha = 0.01$.

c. Do you think the experiment truly examines the difference between the performances of the two brands of SUVs? What variable can influence the results and how can we control for those variables?

d. What assumptions are necessary to make for the validity of the procedures used in this problem?

NOTE: *In the above problem, the sample mean and the sample standard deviation for the two random samples are not given and we have to calculate them. Recall that we learned how to compute sample and sample variance and standard deviation in Chapter 2.*

Exercise 3

A college administrator believes that students who take day classes perform on average better and that their average score on the final exam is above the students who take the same course in the evening by 5 points. To verify this claim, a random sample of 20 students is taken from the day classes of a multisection course and a random sample of 20 is taken from the students taking the same course in the evening. For the day class students, we find that the average score on the final exam was 78.6 with a standard deviation of 8.5 while for the students taking the evening class, we found that the mean was 73.9 with a standard deviation of 9.1.

a. Test to see if the claim of the administrator is valid using $\alpha = 0.5$.

b. Comment on the design of the study.

c. State the necessary assumptions for this problem.

Section 4: The Paired Comparison Test

In the previous two sections of this chapter we learned how to compare the means of two populations in large and small samples. But, every time, we noticed that there could be some other variables that affect the results of experiments and accounting for these variations is often difficult and not always feasible. One way to minimize this variation is to pair experimental units. In other words consider similar experimental units for the two treatments. For example, if we are comparing two methods of teaching calculus to college students, each student in the first group is matched with a student with similar academic background or if we are comparing effectivity of two drugs, we match each patient in the first group with a patient who has similar physical conditions in the second group. In this way, we minimize the effect of other variables. In statistical terminology, this process is called '**blocking**' and experiments with blocking is a very common design in statistical experiments. When blocks consist of two items, we can use the t-test to detect differences in the two populations and the procedure is often referred to as '**matched pair t-test.**'

Suppose that we have n matched pairs. Let $X_1, X_2, \ldots X_n$ be the measurements from the first group and similarly let $Y_1, Y_2, \ldots Y_n$ be the corresponding measurements from the matched individuals in the second group. We consider making inferences on the difference in the measurements for each matched pair. So, let $D_1, D_2, \ldots D_n$ be the differences in the measurements for the two groups, i.e.,

$$D_i = Y_i - X_i \qquad i = 1, 2, X_1, \ldots, n$$

Note that some of the $D_i s$ may be positive and some may be negative. In fact it could happen that some $D_i s$ may even be zero.

Now, we calculate the mean and standard deviation of the differences and denote them by $\overline{D}$ and S_D. It turns out that if we can assume that each sample is taken from a population whose distribution can be characterized by the normal distribution, then the statistic

$$T = \frac{D - \mu_D}{S_D / \sqrt{n}}$$

has a t-distribution with $n - 1$ degrees of freedom. In the above formula, μ_D is the difference in the two population means. Hence we can use the above statistic along with t-distribution to make inferences about the mean difference μ_D in the usual way.

A. Confidence Interval for μ_D

If it is desired to obtain a confidence interval for μ_D with a confidence coefficient of $1 - \alpha$, we determine the critical values $-t_{\alpha/2}$ and $t_{\alpha/2}$ so that the total area under the t-curve between $-t_{\alpha/2}$ and $t_{\alpha/2}$ is $1 - \alpha$. Then the desired confidence interval is given by

$$\overline{X} - t_{\alpha/2} \frac{S_D}{\sqrt{n}} < \mu < \overline{X} + t_{\alpha/2} \frac{S_D}{\sqrt{n}}$$

B. Test of Hypotheses for μ_D

Suppose we wish to test the null hypothesis $H_a: \mu_D = D_0$ against a one- or two-sided hypothesis. Once again, note that in most applications D_0 is zero since we wish to know if the means of the two populations can be assumed to be equal. But, in general, some time we may wish to know if the mean of one population can be assumed to be exceeded by the mean of another population by a fixed amount D_0. Then the test statistic that is used for testing is:

$$T_0 = \frac{\overline{D} - D_0}{S_D / \sqrt{n}}$$

And the value of this statistics is compared to the critical value of the t distribution $n-1$ with degrees of freedom. Thus, as before, if the alternative hypothesis is $H_a: \mu_D < D_0$, then we reject the null hypothesis if $T_0 > t_\alpha$. If the alternative hypothesis is $H_a: \mu_1 - \mu_2 > D_0$, then the null hypothesis is rejected for $T_0 > t_\alpha$ and finally if we have a two-sided alternative $H_a: \mu_1 - \mu_2 \neq D_0$, then we reject H_0 when $T_0 < -t_{\alpha/2}$ or $T_0 > t_{\alpha/2}$. Note that the critical values of the t-distribution t_α and $t_{\alpha/2}$ are based on degrees of freedom.

■ Example 3: Comparing Seasonal Utility Bills

To compare the winter and summer household utility bills in a town, a random sample of 6 homes is selected and their utility bills for the months of February and August are recorded for each home in dollars. The results are given in the table below:

	February Y	August X	Difference D
Home 1	323.08	185.95	137.13
Home 2	198.65	85.91	112.74
Home 3	217.43	112.65	104.78
Home 4	438.19	354.23	83.96
Home 5	285.15	199.4	85.75
Home 6	316.12	201.56	114.56

a. Compute a 95% confidence interval for the mean difference in the utility bills for February and August.

b. Is it reasonable to think that on average the utility bill in February is more than $100 higher than August? Test using $\alpha = 0.5$.

Note that in this problem there are two measurements on each home and therefore there is a natural pairing of each home by itself. It would be erroneous to treat the two samples of utility bills as independent since there are many other factors, e.g., size and efficiency that can affect the utility bill of a household. We use the differences, D in the bills for each home to make inferences. First we compute the mean and standard deviation for the variable D. We have

$$\bar{D} = \$106.49$$

$$S_D = \$19.92$$

Here, the number of degrees of freedom is $n - 1 = 6 - 1 = 5$.

a. Using a *t*-chart or a calculator, we find that the critical value of the t-distribution with 5 degrees of freedom having an area of $\alpha/2 = 0.025$ on each tail is 2.57. Thus for the 95% confidence interval we have

$$106.49 - 2.57 * \frac{19.92}{\sqrt{6}} < \mu_D < 106.49 + 2.57 * \frac{19.92}{\sqrt{6}}$$

$$85.59 < \mu_D < 127.39$$

Hence we are 95% confident that the difference in utility bills for February and August is between $85.59 and $127.39.

b. Here, the null and alternative hypotheses are respectively given by

$$H_0: \mu_D = 100$$

and

$$H_a: \mu_D > 100$$

Therefore the test statistic is calculated as

$$T_0 = \frac{106.49 - 100}{\dfrac{19.92}{\sqrt{6}}} = 0.79$$

which must be compared with $t_{0.05} = 2.015$. Since 0.798 does not exceed the critical value, there is not enough evidence to reject the null hypothesis and we cannot conclude that the mean utility bill for February exceeds that of August by more than \$100.

Exercise 4

An educator wishes to determine if a new teaching strategy has a positive effect of student learning objectives. She administers a pretest to a random sample of 12 students in her class at the beginning of the session and a posttest to the same students upon the conclusion of the course. The scores are given in the table below:

a. Calculate a 95% confidence interval for the mean difference in the final and initial scores of students.

b. Is there evidence to show that the new teaching strategy has been effective in improving the students' scores? Test using $\alpha = 0.5$.

Student Number	Pretest Score	Posttest Score
1	67	80
2	77	88
3	76	69
4	58	71
5	66	79
6	52	62
7	79	95
8	71	86
9	68	84
10	62	78
11	70	74
12	77	85

Exercise 5

Eight people participate in a diet plan for weight reduction for three months. The individuals are weighed at the beginning and the end of the plan. The weights are given in the table below:

a. Determine a 99% confidence interval for the mean weight loss of individuals participating in this plan.

b. Is there evidence to show that on average the weight loss is more than 10 pounds? Test using $\alpha = 0.01$.

Individual	Initial Weight (lb)	Final Weight (lb)
1	192	184
2	221	209
3	186	175
4	236	220
5	188	179
6	208	194
7	214	209
8	228	210